Agricultural Investment and Productivity

Agricultural Investment and Productivity

Building Sustainability in East Africa

EDITED BY

Randall A. Bluffstone and Gunnar Köhlin

Routledge
Taylor & Francis Group
LONDON AND NEW YORK

First published 2011 by RFF Press

2 Park Square, Milton Park, Abingdon, Oxon OX14 4RN
711 Third Avenue, New York, NY 10017, USA

Routledge is an imprint of Taylor & Francis Group, an informa business

First issued in paperback 2017

ISBN: 978-1-61726-091-9 (hbk)
ISBN: 978-1-138-10384-9 (pbk)

Copyedited by John Deever
Typeset by Composition and Design Services
Cover design by Circle Graphics

Library of Congress Cataloging-in-Publication Data

Agricultural investment and productivity : building sustainability in East Africa / edited by Randall A. Bluffstone and Gunnar Köhlin. — 1st ed.
    p. cm.
Includes bibliographical references and index.
ISBN 978-1-61726-091-9 (hb)
    1. Agriculture—Economic aspects—Africa, East. 2. Land use—Environmental aspects—Africa, East. 3. Sustainable agriculture—Africa, East. 4. Agricultural productivity—Africa, East. I. Bluffstone, Randall, 1960- II. Köhlin, Gunnar.
HD2126.A37 2011
338.109676—dc22   2010047939

A catalogue record for this book is available from the British Library

At Earthscan we strive to minimize our environmental impacts and carbon footprint through reducing waste, recycling and offsetting our CO_2 emissions, including those created through publication of this book. For more details of our environmental policy, see www.earthscan.co.uk.

About Resources for the Future *and* RFF Press

Resources for the Future (RFF) improves environmental and natural resource policymaking worldwide through independent social science research of the highest caliber. Founded in 1952, RFF pioneered the application of economics as a tool for developing more effective policy about the use and conservation of natural resources. Its scholars continue to employ social science methods to analyze critical issues concerning pollution control, energy policy, land and water use, hazardous waste, climate change, biodiversity, and the environmental challenges of developing countries.

RFF Press supports the mission of RFF by publishing book-length works that present a broad range of approaches to the study of natural resources and the environment. Its authors and editors include RFF staff, researchers from the larger academic and policy communities, and journalists. Audiences for publications by RFF Press include all of the participants in the policymaking process—scholars, the media, advocacy groups, NGOs, professionals in business and government, and the public. RFF Press is the sister imprint to **Earthscan**, which publishes books and journals about the environment and sustainable development.

About the EfD initiative

The **Environment for Development** (EfD) initiative (www.environmentfordevelopment.org) supports poverty alleviation and sustainable development through the increased use of environmental economics in the policymaking process. The EfD initiative is a capacity-building program focusing on research, policy advice, and teaching.

The EfD is managed by the Environmental Economics Unit of the University of Gothenburg. Financial support is provided by the Swedish International Development Cooperation Agency (Sida). The six EfD centers in Central America, China, Ethiopia, Kenya, South Africa, and Tanzania are hosted by universities or academic institutions in each respective region or country. Resources for the Future and RFF Press are partners in EfD through research collaboration, communications support, and publications, including the EfD book series.

Contents

Contributors

Editors

Randall A. Bluffstone (bluffsto@pdx.edu) is professor of economics and chair of the Department of Economics at Portland State University. His research and teaching interests focus on various aspects of environmental and resource economics, including land use and deforestation in low-income countries, pollution policies in developing and transition economies, environmental liability and privatization and the economics of suburban sprawl. Dr. Bluffstone received his Ph.D. in economics from Boston University in 1993 and from 1983–1985 was a Peace Corps volunteer in Nepal.

Gunnar Köhlin (gunnar.kohlin@economics.gu.se) is associate professor with the Environmental Economics Unit in the Department of Economics at University of Gothenburg and director of the Environment for Development Initiative. As cofounder of the Environmental Economics Unit, he has spent 20 years working with applications of environmental economics in developing countries. His research interests focus on environmental resources, development economics, and in particular the interface between the two. Currently, his research is focusing on sustainable natural resource management in Africa, particularly adoption of soil conservation and the impacts on productivity in Ethiopia.

Contributors

Jean Paul Chavas (jchavas@wisc.edu) is Anderson-Bascom Professor of Agricultural and Applied Economics at the University of Wisconsin, Madison. His recent research focuses on the value of biodiversity, agricultural price analysis, and the economics of farm management. He received his Ph.D. from the University of Missouri, Columbia.

Salvatore Di Falco (s.difalco@lse.ac.uk) is a lecturer in environment and development at the London School of Economics (LSE). Prior to joining LSE, he held a lectureship in applied economics at the Department of Economics at the University of Kent and postdoctoral positions at the National University of Ireland, University of Maryland, and the University of East Anglia. He is an honorary lecturer at the Imperial College London (Wye Campus) and a research fellow with the Environment for Development Initiative (Ethiopia centre). He is associate editor of *Environment and Development Economics*.

Anders Ekbom (anders.ekbom@economics.gu.se) is researcher and lecturer at the Environmental Economics Unit in the Department of Economics at the University of Gothenburg where he heads Sida's External Expert Advice for Environment Economics. His research interests include economic analysis of conservation agriculture, soil erosion, land degradation, and sustainable land management in developing countries. Ekbom holds a Ph.D. in economics from the University of Gothenburg.

Berhanu Gebremedhin (b.gebremedhin@cgiar.org) is a scientist with the International Livestock Research Institute in Addis Ababa, Ethiopia. His research interests focus on policies and institutions for natural and environmental resource management, including impacts of natural and environmental resource technologies, sustainable development of agriculture and commercial transformation of smallholder agriculture. His teaching interest focuses on environmental and natural resource economics, production economics, microeconomics, cost-benefit analysis and applied econometrics. Gebremedhin received his Ph.D. in agricultural economics from Michigan State University.

Fitsum Hagos (f.hagos@cgiar.org) is a researcher with the International Water Management Institute. He has 15 years' teaching and research experience in the economics of agricultural development and natural resource management, including agricultural water management, basin water management, and pollution policies in developing economies. Fitsum received his Ph.D. in development and resource economics in 2003 from the Agricultural University of Norway.

Stein Holden (stein.holden@umb.no) is professor of development and resource economics in the Department of Economics and Resource Management, Norwegian University of Life Sciences. His main research interests are natural resource management and rural development, land tenure, the implications of imperfect markets for behavior, welfare, and development policy. Holden received his Ph.D. from the Norwegian University of Life Sciences (formerly the Agricultural University of Norway) in 1991 and has been professor since 2002. He is coeditor with Keijiro Otsuka and Frank M. Place of *The Emergence of Land Markets in Africa*.

Jane Kabubo-Mariara (jmariara@mail.uonbi.ac.ke) is associate director and senior lecturer in the School of Economics, University of Nairobi. Her research and teaching interests focus on various aspects of development and environment economics including the economics of land conservation and climate change impacts on agriculture and poverty. She received her Ph.D. in economics from the University of Nairobi.

Menale Kassie (m.kassie@cgiar.org) is a scientist at the International Wheat and Maize Improvement Center (CIMMYT) in Nairobi and a research associate with the Environmental Economics Policy Forum at the Ethiopian Development Research Institute in Ethiopia. His research interests focus on development and natural resources economics. Of special interest are the performance of sustainable land management technologies in different agroecological settings and the factors determining sustainable land management technology adoption. Dr. Menale received his Ph.D. in development and resource economics from the Agricultural University of Norway.

Vincent Linderhof (vincent.linderhof@wur.nl) is researcher at the Agricultural Economics Research Institute of the Wageningen University and Research Centre. His research interests focus on the economic aspects of water allocation and the quality of water in low-income and high-income countries, including valuation of water, water policy assessment, and climate change impact assessment. Dr. Linderhof received his M.A. in econometrics from the University of Amsterdam and his Ph.D. in economics from the University of Groningen.

Haileselassie Medhin (hailaat@yahoo.com) is a Ph.D. candidate in the Environmental Economics Unit, University of Gothenburg and a member of EfD-Ethiopia. He completed his M.A. in economics from Gothenburg and holds a B.S. in economics from Addis Ababa University. Haileselassie's research interests include environmental and resource economics and experimental and behavioral economics. He is currently involved in research projects focusing on behavioral aspects of farm households using experimental and survey data from the Ethiopian highlands.

Wilfred Nyangena (nyangena_wilfred@mail.uonbi.ac.ke) is a researcher and lecturer in the School of Economics, University of Nairobi. He also serves as coordinator of the Environment for Development Initiative in Kenya, which is housed within the Kenya Institute for Public Policy and Research Analysis. His research interests include economic analysis of conservation agriculture, soil erosion, land degradation, sustainable land management, and forest reform in developing countries. Nyangena holds a Ph.D. in economics from the University of Gothenburg.

Julius Okello (okelloju@msu.edu) is lecturer and coordinator of the Agribusiness Management Program in the Department of Agricultural Economics at the University of Nairobi. His current research focuses on value chain analysis and the role of information and communications technologies in linking smallholder farmers to high value domestic and international markets.

His other research activities are on safety of peri-urban fresh vegetables, sustainable water management, and sustainable management of community forests. He earned his Ph.D. in agricultural economics from Michigan State University with specializations in international agricultural development, agricultural marketing, and resource and environmental economics.

Bekele Shiferaw (b.shiferaw@cgiar.org) is director of the Socio-Economics Program at the International Wheat and Maize Improvement Center. Previously, he was senior scientist in development and resource economics in Asia and Africa with the International Crops Research Institute for the Semi Arid Tropics. His research interests include institutions and policies for sustainable management and intensification of agro-ecosystems, adoption and impact of agricultural and resource management research, analysis of market relations and value chains, policy and institutional innovations for remedying market failures, and adaptation to and mitigation of climate change impacts on agriculture. Dr. Shiferaw received his Ph.D. in agricultural and resource economics from the Department of Economics and Resource Management, Agricultural University of Norway.

Hailemariam Teklewold (hailemariam.teklewold@economics.gu.se) is a Ph.D. candidate in environmental economics at the University of Gothenburg and Agricultural Economist in Socio-economics with the Ethiopian Institute of Agricultural Research. He is interested in agricultural and resource economics with major emphases on farming systems and sustainable land use, adoption and impact of agricultural technologies, farmers' risk preferences and agricultural investments, agricultural marketing, and value chain analysis.

Mahmud Yesuf (mahmudyesuf@yahoo.com) is a research fellow with Environment for Development Initiative centers in Ethiopia and Kenya. He was also previously an assistant professor of economics at Addis Ababa University. His research and teaching interests focus on environmental and resource economics, including risk aversion and land management in low-income countries, economics of climate change adaptation, and applications of environmental valuation and experimental economics tools in low-income countries. Dr. Yesuf received his Ph.D. in economics in 2004 from the University of Gothenburg.

Acknowledgments

W e would like to thank the contributors to this book who have so generously contributed their time and intellectual energies not only to their own chapters but also to their colleagues' during two authors' workshops held in Ethiopia. These workshops were critical for the preparation of the book, and we would like to thank the Environmental Economics Policy Forum for Ethiopia (EEPFE) at the Ethiopian Development Research Institute (EDRI) for hosting these events. We would especially mention the efforts of Mahmud Yesuf, Alemu Mekonnen and Freweini Berhanu who made the authors' workshops successful. We would also like to acknowledge the important financial support for the authors' workshops and preparation of the book from the Swedish International Development Cooperation Agency (Sida) through the Environment for Development Initiative. Research assistance by Josh Newkirk at Portland State University is also gratefully acknowledged, as is the contribution of Kai Sonder of CIMMYT, who generously contributed the map of the study areas.

The production of the book has benefited greatly from the series editor Thomas Sterner, the close attention given to the project by Don Reisman of RFF Press/Earthscan and the very professional production team at Earthscan.

Randy Bluffstone
Gunnar Köhlin

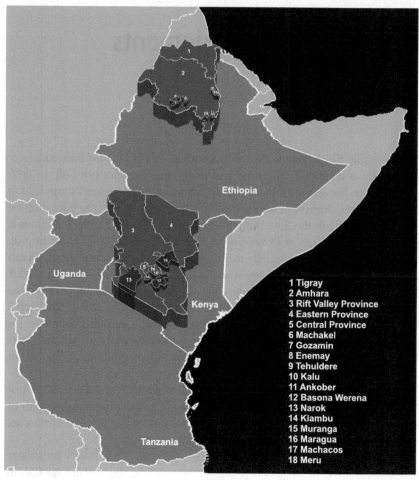

1 Tigray
2 Amhara
3 Rift Valley Province
4 Eastern Province
5 Central Province
6 Machakel
7 Gozamin
8 Enemay
9 Tehuldere
10 Kalu
11 Ankober
12 Basona Werena
13 Narok
14 Kiambu
15 Muranga
16 Maragua
17 Machacos
18 Meru

Study areas discussed in the book

CHAPTER 1

Agricultural Production in East Africa: Stagnation, Investment, and Poverty

RANDALL A. BLUFFSTONE AND GUNNAR KÖHLIN

Agriculture is an enormously important sector in the low-income developing world, because most people live in rural areas and rely on agriculture for their livelihoods. In East Africa, which is the focus of this volume and for our purposes includes Ethiopia, Kenya, Tanzania, and Uganda, over 60% of people live in rural areas. East African people and economies are also some of the most agriculture-dependent in the world. For example, in Ethiopia 84% of households are rural, and similar percentages are farmers. This contrasts with countries such as the United States and Sweden with less than 20% in rural areas and about 2% employed in agriculture. Whereas in Ethiopia 44%, Kenya 28%, Tanzania 46%, and Uganda 32% of GDP come from agriculture, comparable figures in the United States and Sweden are less than 2% (World Bank 2007).

The poor are also concentrated in rural areas. Approximately 90% of the so-called "Bottom Billion" living on less than $1.00 per day live in rural areas. and virtually all of these rural poor rely on agriculture—and particularly subsistence agriculture—for their incomes. Using a more generous poverty line of $2.00 per day we find over two billion rural people who are poor (World Bank 2007; Collier 2007). If only because of its size, agriculture is therefore a critical sector for low-income countries and the poor.

Agricultural growth can be an efficient tool to reduce poverty in Africa (Morris et al. 2007), because so many poor households have few alternatives but to depend on agriculture for their livelihoods. A key reason for this dependence is that with little rural capital formation and land fixed, those involved with agriculture often have only labor as a variable factor of production; agriculture is also typically the sector with the highest proportion of unskilled labor. A number of empirical studies therefore show that GDP growth from agriculture increases consumption among the poorest in the order of three times more than growth originating in the broader economy. One study (Ligon and Sadoulet, reported in de Janvry and Sadoulet 2009) finds that the expenditure effect of agricultural GDP growth is seven times higher for the poorest decile relative to non-agriculture GDP growth.

Unfortunately, the potential of agricultural growth for poverty alleviation has not been realized in sub-Saharan Africa, and as a result poverty has
remained stubbornly high. In 1990 approximately 45% of people lived on less
than $1.00 per day, and as of 2002 the poverty rate was virtually constant at 44%,
though the total number of people in poverty was much higher (ECA 2007).
Unfortunately, the current outlook is bleak and has been so for the past 45 years.
In his address to the Africa Leadership Forum in 1990, then past World Bank
President, Robert S. McNamara noted that in 1983 the Economic Commission
for Africa worried that "The picture that emerges from the analysis of the
perspective of the African region by the year 2008, under the historical trend
scenario is almost a nightmare..." He goes on to note that the situation had
worsened as of 1990, with poverty on the rise, food insecurity, and agricultural
production not keeping up with population growth (McNamara 1990).

Let us use some indicators to exemplify the stagnation in African agriculture mentioned by McNamara and see if we are indeed approaching a
nightmare scenario. The most fundamental task for subsistence agriculture
is to feed the population. Hinderick and Sterkenburg (1983) analyzed agricultural output per capita from 1961 to 1980. While countries such as Côte
d'Ivoire, Swaziland, Malawi, and Rwanda managed improvements, twenty
African countries, including Ethiopia, Tanzania, and Uganda had output per
capita below 1961 levels. And during the period 1990–2004 the picture has in
general worsened. The per capita annual average growth in total food production was −1.2% for Tanzania, −0.6% for both Uganda and Kenya, but +1.8%
for Ethiopia, which is a hopeful development (World Bank 2007).

One reason for the poor per capita production trend is rapid population
growth. In all four countries of interest, populations expanded by anywhere
between 50% and 100% since 1988. Population growth ranged from 2%
per year in Ethiopia to 3.4% annually in Uganda (World Bank 2007), which
highlights the critical role of agricultural productivity in accommodating
population increases. While all other regions have shown a steady increase in
agricultural productivity per hectare and now produce over 5 tons per hectare
in developed countries, 4.5 tons in East Asia, over 3 tons in Latin America,
and 2.5 tons per hectare in South Asia, average productivity in sub-Saharan
Africa has lingered around 1 ton per hectare since 1960. Indeed, in Tanzania
and Kenya average yields per hectare even declined by an average of 0.1% per
year (World Bank 2007).

During this essentially 45-year period of agricultural stagnation or decline
in sub-Saharan Africa every other developing region experienced at least a
doubling of cereal yields. FAO (2007) and Henao and Baanante (2006) show
that Asia increased total cereal output 2.6- to 2.8-fold during the period 1960
to 2004 with increased cultivated area of less than 20%. Africa's cereal production grew by a similar percentage, but in contrast to Asia yields only increased
by 30%, and the increased output required almost a doubling of cultivated land
area. These simple results suggest that to the extent that agricultural production

in sub-Saharan Africa has kept up with population growth, it has done so by increasing cultivated area rather than through higher yields per hectare.

After decades of decline, agricultural livelihoods and productivity in sub-Saharan Africa are approaching critical levels. Millions of people in East Africa simultaneously rely on agriculture for their sustenance and live in abject poverty. Indeed, in many countries achieving the Millennium Development Goal to halve poverty by 2015 remains a dream (Bluffstone et al. 2008), mainly because rural incomes tied to agriculture are so low. Yet the way forward is unclear, at least partly because relatively little is known about the factors promoting and inhibiting the investments necessary to improve agriculture.

Explaining Agricultural Stagnation in Africa

An important reason for stubbornly high levels of extreme poverty, low output levels, and yield stagnation is technology stagnation leading to declining soil quality. Most of sub-Saharan Africa experiences substantial soil fertility losses every year. During the 2002–2004 cropping period 85% of African agricultural lands or about 185 million hectares had nitrogen, phosphorus, and potassium nutrient losses of over 30 kilograms per hectare per year (kg/ha/yr) and 40% lose more than 60 kg/ha/yr. Over time these losses have resulted in about 95 million hectares becoming virtually uncultivable, largely because of soil erosion and leaching. If this trend continues, agricultural yields could be reduced 17% to 30% by 2020 (Henao and Baanante 2006).

As shown in Table 1-1, East African agriculture is very much a part of this disappointing trend. All countries considered in this volume experienced

TABLE 1-1 Estimated Average Nitrogen, Phosphorus, and Potassium Soil Nutrient Losses in Select African Countries, 2002–2004 Cropping Year

Moderate/Low *< 30 kg/ha/yr*		*Medium* *30 to 60 kg/ha/yr*		*High* *> 60 kg/ha/yr*	
Country	*Loss (kg/ha/yr)*	*Country*	*Loss (kg/ha/yr)*	*Country*	*Loss (kg/ha/yr)*
Egypt	9	Swaziland	37	Tanzania	61
Mauritius	15	Burkina Faso	43	Mauritania	63
South Africa	23	Botswana	47	Lesotho	65
Zambia	25	Sudan	47	Madagascar	65
Morocco	27	Ethiopia	49	Uganda	66
Algeria	28	Mozambique	51	Kenya	68
		Eritrea	58	Malawi	72
		Ghana	58	Somalia	88

Source: Morris et al. (2007) quoting Henao and Baanante (2006)

4

at least medium fertility losses and, due to reasons such as soil erosion, three of four lost more than 60 kg/ha/yr on average (Morris et al. 2007). Yesuf et al (2005) indeed point out that nutrient losses due to land degradation are a significant drag on productivity. From a review of studies they conclude that in Ethiopia land degradation on average reduces agricultural productivity by an average of 2 to 3% per year. Indeed, land degradation probably explains the stagnation of cereal yields in Ethiopia (Morris et al. 2007).

In the case of East Africa the challenge is therefore, at least in part, to develop appropriate technologies that stem these fertility losses and are profitable for and will actually be adopted by farmers. Such technologies must also be sustainable given the agro-ecological conditions.

That improved technologies can have impressive effects on productivity has been shown in other regions. Implementation of green revolution technologies in China produced about half the gains in rice production during the period from 1975 to 1990 (World Bank 2007). Experiences in Africa have been less impressive. The Consultative Group for International Agricultural Research (CGIAR) African Rice Initiative (2008) notes that the adoption of NERICA rice varieties led to an increase in rice production in West Africa, with Guinea leading the way, with a humble 5% increase.

But better practices can have huge effects. Ethiopian field research suggests that using advanced technologies can yield 5.1 tons of grain per hectare, and similar results were found in Uganda, where yields jumped from 1.8 to 4.2 tons per hectare. These yields are similar to those in China and Vietnam (CGIAR 2008).

Agricultural research and development is a key driver of new agricultural technologies, but in general agricultural R&D spending has been very low. In 2000, for example, spending across 27 sub-Saharan African countries was about $1.5 billion 1993 dollars, with about one half of $1 billion spent in East Africa. This level of sub-Saharan Africa expenditure is above the 1991 total of about $1.1 billion, but it is pitifully low given the economic and food security challenges the continent faces (Beintema and Stads 2004).

Agricultural R&D spending grew at a 6.8% average annual rate during the 1960s, but since 1971 growth has been low or negative. During the 1980s spending grew 1% annually, but at the start of the 1990s the positive growth in R&D turned negative for half of the 27 sub-Saharan African countries. Excluding South Africa and Nigeria, agricultural R&D spending in sub-Saharan Africa declined by an average of 0.2% per year during the 1990s (Pardey et al. 1997; Beintema and Stads 2004).

A common measure of agricultural research effort is research intensity, which measures research expenditures as a percentage of agricultural GDP. This ratio fell by 26% in real terms during the period 1981 to 2000, from $0.95 to $0.70 spent on research per $100 of output. The 2000 East African average is $0.50 compared with a developing country average of $0.62 and a sub-Saharan African average of $0.79. Ethiopia averaged 0.25%, Tanzania

0.3%, and Kenya 1.1% of agricultural GDP. The InterAcademy Council (www. interacademycouncil.net) in 2004 found this level of commitment wanting and recommended that Africa's agricultural research-intensity ratio double by 2015. Achieving this goal would require expenditure growth of 10% annually for a decade (Beintema and Stads 2004). With the mentioned commitments from the African governments, G8, and also the active participation of private foundations, such as the Gates Foundation, such an expansion is within reach.

Government spending on and overseas development assistance (ODA) to agriculture are broader indicators of the importance given to agriculture as an engine of growth and poverty alleviation. In Ethiopia government spending in 2004 accounted for 4.3% of agricultural value added while 6.4% of ODA was directed to agriculture. For both Kenya and Uganda government spending was 4.1% while ODA to agriculture was three times higher in Kenya than Uganda (10.4% vs. 3.5%). Tanzania had no reported government spending in 2004, but 5.7% of ODA went to agriculture (World Bank 2007). Most significantly, the World Bank reports that ODA spending on agriculture fell dramatically in all four East African countries in the early 1980s and also in Ethiopia and Tanzania in the period 2000–2004.

Chemical fertilizers have provided considerable growth in agricultural production throughout the world, and the World Bank credits increased use of fertilizers with approximately 20% of the agricultural growth that has occurred over the last 30 years in developing countries. In Asia chemical fertilizer use skyrocketed from an average of 6 kilograms per hectare in 1961–1963 to over 100 kilograms per hectare in 2000–2002. Average annual applications in China are estimated to be 395 kilograms and in South Korea 389 kilograms per hectare.

Table 1-2 presents average annual fertilizer intensity and growth by developing region. We see in the table that fertilizer use grew throughout the world during the period 1962 to 2002. Applications started out at very low levels in South Asia and particularly in sub-Saharan Africa, but they grew very slowly in Africa. By 2002 average fertilizer intensity in sub-Saharan Africa was estimated at only 8 kg/ha/yr or only an eightfold increase over 40 years. By contrast, fertilizer use in South Asia increased by over 33-fold; during that

TABLE 1-2 Average Fertilizer Use Intensity and Annual Growth by Developing Area

Region	Average Use (kg/ha/yr)			Average Annual Growth	
	1962	1982	2002	1962–1982	1982–2002
South Asia	3	38	101	13.2%	5.0%
East and S.E. Asia	12	53	96	7.6%	3.4%
Latin America	10	43	78	7.8%	3.1%
Sub-Saharan Africa	1	7	8	8.7%	0.9%
All Developing Countries	6	52	102	11.3%	2.3%

Source: Morris et al. (2007)

period the region dramatically stabilized its food production. Much of the slowdown in growth of fertilizer use occurred since 1982. Indeed, during the 1990s fertilizer intensity actually declined. Currently, sub-Saharan Africa uses less than 1% of the world's chemical fertilizer (Morris et al. 2007).

Average applications of fertilizer in East Africa are virtually the same as the rest of the continent (Morris et al. 2007). Estimates differ depending on the source, but the message is clear that with the exception of Kenya, fertilizer intensity is astoundingly low. Kenya uses 32 to 44 kg/ha/yr, which is a fraction of the developing country average, but substantially higher than the rest of the region. By contrast, Tanzania applies 5 to 13 kg/ha/yr, Ethiopia uses 3 to 14, and the Uganda intensity averages 0.6 to 1.0 kg/ha/yr (Crawford et al. 2006; World Bank 2007).

Irrigation is also critical. Across the developing world the percentage of crops aided by irrigation grew throughout the 1990s, reaching 39% in South Asia and near 30% in Latin American countries. The sub-Saharan Africa average is only 4%, and in the last 40 years sub-Saharan Africa has irrigated only 4 million hectares of land, the lowest of all regions. In our four countries of interest the top irrigator is Tanzania (3.5% of cultivated land), while the lowest is Uganda (0.1%) (World Bank 2007). UNEP (2008) estimates that by 2050 nearly a third of sub-Saharan Africa will suffer water shortages, which suggests that irrigation will only become more important over time.

Soil structure and fertility are also serious problems in sub-Saharan Africa. UNEP (2008) illustrates the state of much farm land in sub-Saharan Africa. Only 10% of farm soil is rated as "prime," while 25% falls into the "moderate" or "low" sustainability categories. Erosion and damage from chemicals has caused 65% of farm land to suffer some form of degradation. Ethiopia is estimated to lose 3% of agricultural GDP each year due to soil degradation (World Bank 2007).

Sustainable land management (SLM) technologies like stone walls, check dams, furrow plowing, and soil bunds can reduce this degradation and are likely insufficiently used. According to the International Fertilizer Development Center, sub-Saharan Africa loses 52 kilograms of nitrogen-phosphorus-potassium per hectare every year, which is five times the gain from fertilizer.

Opportunities and Challenges in Agricultural Technology Adoption

Reducing erosion and increasing productivity are not nearly as simple as just putting guaranteed-appropriate technologies in farmers' hands, and this is perhaps the key point of many chapters in this volume. Not all technologies are appropriate for all farmers, but which ones increase output and profitability may not be obvious. Indeed, Nkonya et al. (2008) show that Kenyan farmers who undertake all available SLM practices incur losses equivalent to one-third of household incomes! Farmers who stagger implementation still suffer losses, but they are much smaller; even if SLM technologies pay off eventually, with very high up-front investment costs and limited or nonexistent credit markets it is perhaps not surprising that African farmers can be slow to adopt. Recognition of landscape-level benefits of SLM like reduced off-site erosion may provide rationales for offering financial assistance and improving social capital to create economies of scale.

In a comprehensive review of the literature on land management in the Ethiopian highlands, Yesuf and Pender (2005) find that many factors impede adoption, including poverty, high investment costs, limited access to agricultural inputs and credit, low returns to agriculture and many conservation practices, risks, insecure land tenure, short time horizons of farmers, and lack of information about appropriate technologies. Many of these factors are affected by policies toward infrastructure and market development, land tenure, agricultural research and extension, conservation programs, land use regulation, and local governance and collective action.

There is now a renewed interest in agriculture as a vehicle for development (World Bank 2007; de Janvry and Sadoulet 2009). The challenge will then be to design interventions that stimulate pro-poor growth in the agricultural sector. In the wake of the "food crisis" that preceded the 2008–2009 worldwide financial crisis, investments in agriculture are finally back on the international agenda. FAO calculates that developing country agriculture needs $30 billion a year in investment to help farmers achieve the goal of reducing the number of hungry people in half by 2015 (FAO 2009).

At the L'Aquila Summit, the G8 committed $20 billion over three years to support sustainable agricultural development (G8 2009). This will partly be implemented through the Comprehensive African Agriculture Development Programme that is intended to eliminate hunger and reduce poverty through agriculture. African governments have agreed to increase public investment in agriculture by a minimum of 10% of their national budgets and to increase agricultural productivity by at least 6% (CAADP 2010). These are far from the first such initiatives and pledges, so the challenge for policymakers and donors will be to make these proposed investments productive.

This book focuses primarily on SLM investments such as erosion control, though other potentially productivity-enhancing investments such as more appropriate seed varieties and chemical fertilizers are also considered. SLM is a key focus, because to prevent erosion and loss of soil capital, critical problems in East Africa, SLM is one of the most important, costly, and durable investments East African farmers make. It is therefore of critical importance to understand the determinants, returns, and public policies affecting adoption. The focus of the book is to understand adoption from the perspective of the farmers, probing into the determinants of investments (Part I), how investment decisions are affected by risk (Part II), how to analyze the returns from investments (Part III), and finally how farmers' decisions might be affected by public policies (Part IV).

Overview and Contribution of the Volume

Part I looks at the determinants of investments, beginning with a chapter by Bekele Shiferaw and Julius Okello that provides a modern framework for analyzing agricultural investments. The chapter particularly focuses on SLM investments. Wilfred Nyangena then examines the social aspects of SLM investments in Kenya. Using principle components analysis and random effects probit models, he finds that social capital represented by strong social arrangements can encourage investments in SLM. This is likely due to the nature of the SLM, which has an amazing array of public, club, and private good characteristics.

The final chapter in Part I by Jane Kabubo-Mariara and Vincent Linderhof examines the role of tenure security in grass strip and terracing investments in Kenya. Using probit models combined with factor analysis, they find that household plots with strong property rights are significantly more likely to exhibit investments. With the exception of variables found by other researchers to be important, such as market access and agro-ecological potential, other factors typically offer inconclusive results, suggesting that tenure security is particularly important.

Part II focuses on how risk affects investments. Mahmud Yesuf and Hailemariam Teklewold examine risk aversion using experimental data from real payoff games and find that a majority of respondents in three zones in highland Ethiopia exhibit severe or extreme risk aversion; this level of risk aversion is higher than similar studies in India and Zambia. They then examine the determinants of risk aversion and find that because of missing markets wealth is a critical determinant. Using two-stage random effects Heckman selection models the authors analyze the relationship between risk aversion and fertilizer use, which is very low in Ethiopia. They find that depending on the study site, risk aversion is negatively and significantly associated with fertilizer use and/or application intensity. This suggests that a key part of promoting fertilizer use is insuring households against losses.

Fitsum Hagos and Stein Holden also examine the relationship between fertilizer use and intensity and risk aversion. Using experimental data on risk aversion from the highlands of Tigray Regional State in Ethiopia, they find results that contradict those of Yesuf and Teklewold. Indeed, because the existence of chemical fertilizer use is positively and statistically significantly related to risk aversion, they conclude that chemical fertilizer—while risky—must enhance food security for the over 90% of households who are net buyers.

Salvatore Di Falco and Jean-Paul Chavas use a production function approach to examine the relationship between crop biodiversity and risk as measured by the first three moments of the production function distribution. Biodiversity is measured using indices related to numbers of barley and cereal varieties employed per unit of area with a particular focus on indigenous "landraces" developed by farmers. Using data from Tigray in Ethiopia they find that biodiversity increases mean yields and also helps farmers manage risk. An innovation of this chapter is to note the important difference between variance, which does not distinguish between upside and downside risk, and skewness, which can be interpreted as the probability of very low or very high outcomes. They find that biodiversity increases variance, but also increases skewness, reducing the probability of crop failures. They conclude that, on balance, biodiversity reduces production risk.

Part III includes two chapters that focus our attention on the returns to agricultural investments. Menale Kassie asks whether and under what circumstances two key SLM investments are profitable for farmers in the Ethiopian highland Amhara and Tigray Regional states. Using propensity score matching and switching regression techniques applied to plot-level data he finds that whether reduced tillage and stone bunds are profitable for farmers critically depends on the state in which plots are located. Indeed, these SLM investments yield high returns in Tigray, but not in Amhara Regional State. As Amhara has approximately three times the average annual rainfall of Tigray, he concludes that because SLM investments conserve *both* soil and water, they are likely to be most beneficial in lower rainfall areas.

The chapter by Haileselassie Medhin and Gunnar Köhlin examines the efficiency of SLM investments. In a theoretical model they decompose efficiency into technical and allocative efficiency and remind us that because of heterogeneous technologies farmers may operate on their technological frontiers, but at very different production levels. Using plot-level data from highland Ethiopia, they test whether plots with four different SLM technological profiles represent fundamentally different production functions and find that pooled estimation is not justified. They examine returns and find that mean productivity is highest on plots without SLM investments, which raises the question of whether SLM perhaps represents a failed set of technologies. Digging deeper, they examine performance under different conditions and find that SLM investments tend to be made on plots with steeper slopes and lower soil qualities. They also find that plots with SLM are overrepresented in the pool of best-performing plots, leading them to conclude that SLM technologies help disadvantaged plots increase their productivity and perform more similarly to inherently more productive plots than would have otherwise been the case.

Part IV explores public policy options. Anders Ekbom examines the issue theoretically and argues that the critical public policy issue is to create incentives to internalize off-site effects of soil erosion and runoff. He particularly focuses attention on information and extension, charges and fees (which he concludes would be difficult to implement), improved land tenure, and payments for environmental services (PES). He concludes that PES schemes are especially promising, because they reward farmers for socially beneficial activities like SLM and also charge downstream users for those services.

The penultimate chapter presents a comprehensive and wide-ranging framework for categorizing instruments for SLM and evaluating East African policies. In this chapter Berhanu Gebremedhin argues that government policies

to promote SLM can affect farmer activities directly (e.g., financial support
for SLM investments) or influence their decisions indirectly (e.g., tax relief or
crop price supports). Another policy dimension is whether laws and regulations augment private efforts in the name of improving overall social welfare or
enabling investments. For example, improving market mechanisms (e.g., credit
markets) would be enabling. Policies may then be direct and augmenting (e.g.,
subsidies), direct and enabling (e.g., targeted training), indirect and augmenting
(e.g., land tenure reform) or indirect and enabling (e.g., promotion of collective
action). He then evaluates the policies used in Ethiopia, Kenya, Tanzania, and
Uganda, concluding that in general indirect incentives have proved to be most
successful. The final chapter concludes the volume and draws out key lessons.

Acknowledgment

The authors are grateful for expert research assistance from Joshua Newkirk.

References

Beintema, N., and G. J. Stads. 2004. Sub-Saharan African Agricultural Research: Recent
    Investment Trends. *Outlook on Agriculture* 33(4): 239–46.
Bluffstone, R., M. Yesuf, B. Bushie, and D. Damite. 2008. Rural Livelihoods, Poverty,
    and Achieving the Millennium Development Goals in Ethiopia. *EfD Discussion
    Paper* 08–07. Washington DC: Resources for the Future.
CAADP (Comprehensive Africa Agriculture Development Programme). 2010. www.
    nepad-caadp.net (accessed January 4, 2010).
CGIAR (Consultative Group on International Agricultural Research). 2008. African
    Rice Initiative Regional Coordination Unit and the Countries of the Multinational
    NERICA Rice Diffusion Project. Washington, DC: CGIAR.
Collier, P. 2007. *The Bottom Billion*. Oxford: Oxford University Press.
Crawford, E., T. S. Jayne, and V. Kelly. 2006. Alternative Approaches for Promoting
    Fertilizer Use in Africa. *Agriculture and Rural Development Discussion Paper* No. 22.
    Washington, DC: World Bank.
de Janvry, A., and E. Sadoulet. 2009. Agricultural Growth and Poverty Reduction:
    Additional Evidence. *The World Bank Research Observer* 25(1). Advance Access
    November 9, 2009.
Economic Commission for Africa (ECA). 2007. *Economic Report on Africa 2007:
    Accelerating Africa's Development through Diversification*. Addis Ababa, Ethiopia:
    United Nations Economic Commission for Africa.
FAO (United Nations Food and Agricultural Organization). 2007. *The State of Food
    and Agriculture: Paying Farmers for Environmental Services*. Rome, Italy: Electronic
    Publishing Policy and Support Branch, FAO.
———. 2009. *Achieving Food Security in Times of Crisis*. Rome: FAO.
G8 (Group of Eight). 2009. "L'Aquila" Joint Statement on Global Food Security, July
    10, 2009. www.g8italia2009.it/static/G8_Allegato/LAquila_Joint_Statement_on_
    Global_Food_Security%5B1%5D,0.pdf (accessed January 4, 2010).

Henao, J., and C. Baanante. 2006. "Agricultural Production and Soil Nutrient Mining in Africa: Implications for Resource Conservation and Policy Development Summary." International Center for Soil Fertility and Agricultural Development (IFDC), Muscle Shoals, Alabama. March. Accessed April 29, 2010. www.eurekalert. org/africasoil/report/Soil_Nutrient_Mining_in_Africa_Report_Final.pdf

Hinderick, J., and J. J. Sterkenburg. 1983. Agricultural Policy and Production in Africa: the Aims, the Methods, and the Means. *The Journal of Modern African Studies* 21(1): 1–23.

InterAcademy Council. 2004. *Realizing the Promise and Potential of African Agriculture: Science and Technology Strategies for Improving Agricultural Productivity and Food Security in Africa*. Amsterdam, the Netherlands: IAC Secretariat.

McNamara, R. 1990. "Africa's Development Crisis: Agricultural Stagnation, Population Explosion, and Environmental Degradation." Address to the Africa Leadership Forum, Ota, Nigeria. June 21, 1990. Washington DC: World Bank.

Morris, M., V. Kelly, R. Kopicki, and D. Byerlee. 2007. *Fertilizer Use in African Agriculture: Lessons Learned and Good Practice Guidelines*. Washington, DC: The World Bank.

Nkonya, E., P. Gicheru, J. Woelcke, B. Okoba, D. Kilambya, and L. N. Gachimbi. 2008. On-Site and Off-Site Long-Term Economic Impacts of Soil Fertility Management Practices. *IFPRI Discussion Paper* 00778. Washington, DC: International Food Policy Research Institute.

Pardey, P., J. Roseboom, and N. Beintema. 1997. Investments in African Agricultural Research. *World Development* Vol. 25(3): 409–423.

UNEP (United Nations Environment Programme). 2008. *Africa: Atlas of Our Changing World*. Nairobi, Kenya: Division of Early Warning and Assessment, UNEP.

World Bank. 2007. *World Development Report 2008: Agriculture for Development*. Washington DC: The World Bank.

Yesuf, M., A. Mekonnen, M. Kassie, and J. Pender. 2005. Cost of Land Degradation in Ethiopia: A Critical Review of Past Studies. *Environment for Development Report*. www.efdinitiative.org/research/projects/project-repository/economic-sector-work-on-poverty-and-land-degradation-in-ethiopia (accessed January 26, 2011)

Yesuf, M., and J. Pender. 2005. Determinants and Impacts of Land Management Technologies in the Ethiopian Highlands: A Literature Review. *Environment for Development Report*. www.efdinitiative.org/research/projects/project-repository/ economic-sector-work-on-poverty-and-land-degradation-in-ethiopia (accessed January 26, 2011)

Determinants of Sustainable Land Management Investments

PART I

Determinants of Sustainable
Land Management
Investments

CHAPTER 2

Stimulating Smallholder Investments in Sustainable Land Management: Overcoming Market, Policy, and Institutional Challenges

BEKELE SHIFERAW AND JULIUS OKELLO

The degradation of natural resources raises a variety of issues related to rural livelihoods, poverty, distribution of income, and intergenerational equity. Land degradation also deprives smallholders and particularly the poor of a key resource and diminishes capacity to undertake critical investments, possibly leading to depletion of buffer stocks and increased vulnerability. These problems are most pronounced in areas with widespread poverty and fragile ecosystems, such as arid, semi-arid and highland regions (Pender and Hazell 2000; Shiferaw and Bantilan 2004). In such areas, sustainable intensification of agriculture through land conservation and management is a critical policy challenge.

In recognition of the importance of land degradation for rural livelihoods, governments and development partners in East Africa have devoted substantial resources to developing and promoting soil and water conservation technologies. These methods are diverse and include both indigenous and introduced practices for combating soil erosion and nutrient depletion, improving water conservation, and enhancing productivity. Structural methods are often promoted through donor-financed projects (e.g., food for work) and include soil or stone bunds and terraces. Agronomic practices include minimum tillage, organic and inorganic fertilizers, grass strips and agro-forestry. These techniques aim to reduce soil erosion while increasing organic matter and increasing nitrogen fixation. In addition, water harvesting techniques like tied-ridges, planting basins, check-dams, ponds, tanks, and bore wells provide farmers the opportunity to plant early, better utilize available moisture for plant growth, and reduce reliance on unpredictable rains (Baidu-Forson 1999).

Despite the growing policy interest, widespread adoption of sustainable management techniques outside of intensively supported projects has been

15

limited (Fujisaka 1994; Pender and Kerr 1998; Barrett et al. 2002). A review of the literature suggests that, while there is still inadequate understanding of the role of market, policy, and institutional factors in shaping incentives for adoption, market and policy failures can create important barriers to smallholder adoption of sustainable land management (SLM) (Zaal and Oostendorp 2002).

A number of chapters in this volume amplify this point and contribute to this literature. In Chapter 4, Kabubo-Mariara and Linderhof, for example, find that secure land tenure, which in East Africa is largely a function of the policy environment, is critical for farmer investments. Nyangena in Chapter 3 broadens the point that institutional factors are critical. While also finding that tenure security promotes SLM investment, he concludes that various types of social capital are critical for adoption. Social relations of various types therefore appear to be important for investments.

This chapter reviews the challenges smallholder farmers face in tackling the long-standing problem of land degradation and offers new insights into how market incentives, institutional factors, and macroeconomic policies affect adoption and adaptation of land and water management technologies. The chapter is organized as follows. The following section discusses the evolution of approaches to soil and water conservation. The next section provides a broad conceptual framework for analysis and evaluates challenges. A review of factors that condition the use of sustainable land and water management follows, and the final section offers conclusions, key lessons, and implications for policy and future research.

Evolution of Approaches for Sustainable Land and Water Management

Concern with land and water degradation is not new. Over the years considerable effort has gone into getting smallholder farmers to mitigate land degradation and adapt existing techniques to local conditions. Reducing soil erosion and associated nutrient depletion has been a particular priority, and due to off-site effects like siltation of reservoirs and waterways, governments have often intervened to reduce soil erosion and runoff in hilly areas. In semi-arid regions the focus is often on capturing and utilizing surface and groundwater. Most efforts have met with limited success.[1]

Promotion approaches can be grouped into top-down, populist or "farmer-first," and neoliberal (Biot et al. 1995). Most of the land management interventions by colonial governments were top-down command-and-control type policies that did not involve smallholder farmers and were driven by fear of inaction. Policies included forced adoption of erosion control, planting of trees, and protection of water/river catchments. Until the mid-1980s several countries in East Africa used similar policies (e.g., see Shiferaw and Holden 1998; Pandey 2001). These approaches largely failed and created serious barriers to innovation.

The failure of command-and-control led to the so-called "populist" approach, which largely rejected external technology development and extension and made the farmer the center of soil and water conservation programs. Chambers et al. (1989) is emblematic of this approach, stressing small-scale, bottom-up interventions, often using indigenous technologies (Reij 1991). Although the idea of putting farmers first is noble, implementation proved difficult, leading to a broader approach in which farmer innovation is affected by economic, institutional, and policy environments (Biot et al. 1995; Robbins and Williams 2005).

The neoliberal approach focuses on incentives that condition the use of land and water management technologies. This framework recognizes the role of farmer innovation, but highlights the critical influence of markets, policies, and institutions for farmer innovation, adoption, and adaptation. The critical importance of making conservation attractive and economically rewarding to farmers through productive technologies and access to markets are regarded as key to success.

Growing recognition of the public good characteristics of soil and water conservation and the nontechnical factors that condition technology choice have led to strategies that internalize local externalities at the community and landscape levels (Pagiola 1998; Reddy 2005; Kerr et al. 2007). Soil conservation provides off-site benefits that include better water quality and flood control for downstream users (Ribaudo 1986; Fox et al. 1995; Colombo et al. 2006). Integrated watershed management (IWM) aims to improve both private and communal livelihoods through technological and institutional interventions. IWM goes beyond traditional soil and water conservation to include collective action, networking, and market-related innovations that support and diversify livelihoods. This concept ties together the watershed with community and institutional factors that determine viability and sustainability. Linking the biophysical concept of a watershed with communities can help develop technologies and local collective action to internalize externalities and stimulate investments that address community-wide resource management problems (Shiferaw et al. 2008a).

In the last few years soil and water conservation has recognized design complexities and the need for broadening partnerships and disciplinary analyses and has moved toward sustainable land and water management (Robbins and Williams 2005). There is no single definition for SLM. Hurni (2000) suggests that SLM implies "a system of technologies and/or planning that aims to integrate ecological, socioeconomic, and political principles in the management of land for agricultural and other purposes to achieve intra- and intergenerational equity." The following section builds on this concept of SLM and develops a conceptual framework for understanding the market, policy, and institutional factors that affect investment in conservation. Understanding the drivers of these decisions will allow the design of win-win SLM strategies that reduce poverty and increase agricultural output.

18   *Bekele Shiferaw and Julius Okello*

Conceptual Framework

Small farmers in many developing regions produce and consume the same commodities, which means that investments in land and water management are likely to be influenced by factors related to both production and consumption. This is especially true when farmers operate under imperfect information and market conditions that prevent them from producing for sale and profits. Our framework presented in Figure 2-1 presumes that farm households pursue livelihood strategies constrained by a variety of factors as they make decisions about natural resources and investments. The framework is premised on Chambers (1987) and the farmer-first principles, but it also incorporates farm household behavior under market imperfections (de Janvry et al. 1991), economics of rural organization (Hoff et al. 1993), economic policies (Heath and Binswanger 1996), and institutional economics (North 1990).

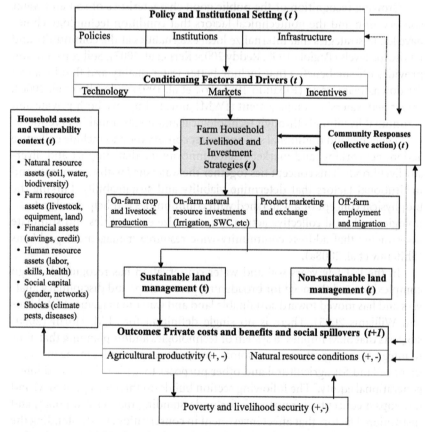

FIGURE 2-1 Factors Conditioning Smallholder Natural Resource Investments and Development Pathways

Smallholder farmers make production and investment decisions in each period to maximize net benefits, subject to existing assets and expected shocks. These two factors determine vulnerability. Decisions are affected by socioeconomic and policy environments, institutional changes, and infrastructure that determine relative prices and access to technologies and markets (Shiferaw and Bantilan 2004). Market access is further influenced by information imperfections and the high search costs that prevail in many developing countries (Fafchamps and Hill 2005). Institutional factors affect sustainable land and water management through legal frameworks, property rights, and farmer participation in networks. In cases like watershed management, collective action may support production and investments.

Household assets and the prevailing biophysical, socioeconomic, and institutional environments jointly determine the livelihood options and investment strategies available to farmers. Access to input and output markets, technologies, and the resulting prices then define the feasible production set and determine the optimal investment strategies. Enabling and efficient institutions (e.g., secure rights to land and water and functioning credit and extension systems) also support investments that provide opportunities to intensify production, diversify livelihood strategies, and potentially combat resource degradation.

The interplay of technological and institutional factors can spur households to pursue potentially sustainable intensification that improves livelihoods. In the absence of enabling policy and institutional environments that encourage technological innovation, farmers lack the incentives to use SLM technologies. Indeed, lack of viable technological options and adverse biophysical, policy, and institutional environments can encourage exploitative and unsustainable livelihood strategies, leading to synergies between poverty and resource degradation and potentially downward spirals (Scherr 2000).

Efficient use of SLM is also affected by its nature as a public good. The costs of conserving land and water are paid by investors, but the benefits accrue to agents well beyond the farm. Significant off-site SLM benefits present challenges, because they can lead to underinvestment. This is particularly true when, as is often the case, the effectiveness of conservation investments depends on treating an entire catchment or micro-watershed (as Ekbom emphasizes in Chapter 10); this requires collective action and landscape-wide cooperation, but such cooperation often involves costs and leads to additional market failures. These issues are discussed further in the following section.

Determinants of Conservation Investments

Investment in SLM is often just one of many investment options available to farmers. One way to model behavior is to suppose farmers compare the expected costs and benefits of all options and invest in those that offer the highest net returns (Kerr and Sanghi 1992; Pagiola 1998; Lee 2005); farmers therefore switch from old to new methods when they gain in terms of net returns, lower risks, or both. Particularly with large off-site benefits, highest private returns might come from investments other than soil and water conservation; adoption will therefore be inhibited unless subsidies are offered.

The conceptual framework presented in the previous section identifies factors that condition the adoption and adaptation of soil and water management interventions in smallholder agriculture. In the context of Figure 2-1, in addition to environmental factors determinants can broadly be categorized as policy/institutional and market, poverty, and risk. These are discussed below.

Agricultural Policy and Institutional Factors

In the past decade there has been an increasing recognition that policy and institutional arrangements play important roles in sustainable management of natural resources (Heath and Binswanger 1996; Barbier 2000; Pandey 2001; Zaal and Oostendorp 2002; Reddy 2005). We focus on some of the most direct influences of agricultural policies on SLM investments. Though there is a movement to reintroduce some targeted subsidies for fertilizer, seeds, and irrigation (Kelly et al. 2003), unlike in some Asian countries (e.g., India), most countries in sub-Saharan Africa (except Malawi) have done away with agricultural input and investment subsidies. Public support for irrigation water and infrastructure is an important example.[2] In India, as in many Asian countries, irrigation water is typically free and electricity subsidized (Reddy 2005). These policies distort incentives and can create disincentives for investment in soil erosion control and conservation of available water (Reddy 2005; Shiferaw et al. 2008a). They can also encourage the planting of water-intensive crops, often in semi-arid regions, and SLM investments may be short-lived as farmers resort to old practices once subsidies are withdrawn. While subsidies can be justified by market and institutional failures, such policies must be carefully appraised.

Institutions are the rules, enforcement mechanisms, and organizations that help shape expectations and behavior and facilitate market and non-market transactions. They transmit information, mediate transactions, facilitate collective action, regulate property rights and contracts, and help internalize externalities. As discussed in the literature and a number of chapters in this volume, of special importance for SLM are property rights, collective action, and social networks.

Access and security of rights to land, water, and other natural resources are important, because if property rights are weak farmers cannot capture the full benefits of their investments; therefore, incentives to invest in SLM may be reduced (Ahuja 1998; Barrett et al. 2002; Shiferaw and Bantilan 2004). However, empirical evidence on the effect of land ownership rights on SLM is mixed. Knowler and Bradshaw (2007) review thirteen studies that assess the impact of land ownership on adoption of SLM in several countries. They find that in two cases owned land is better maintained, but in three cases the opposite is found, and in the rest no relationship exists.

As was already discussed, two chapters in this volume find that tenure security promotes investments. This point is also emphasized in Chapter 11 by Berhanu Gebremedhin, who discusses incentives for SLM. Relying on his and others' previous empirical work, he determines that tenure security is an essential underpinning for introducing a variety of incentives.

When SLM provides important flood and soil erosion control in community watersheds there are public goods externalities, and incentives for private investments may be limited. In such cases interdependence of resource users will require collective action and cooperation to achieve socially desirable conservation outcomes. Evidence suggests policies and institutions that induce and sustain collective action can play a significant role in the conservation and management of communal resources. Ahuja (1998) and Gebremedhin et al. (2003) examine the effects of collective action on adoption of conservation technologies in Côte d'Ivoire and Ethiopia and find—as does Nyangena for Kenya in Chapter 3—that collective action supports adoption of conservation practices by helping farmers to address market failures and overcome information constraints.

Networking among farmers, including participation in the design of land management technologies, has an important role in influencing farmers' attitudes and perceptions. Networking facilitates access to information about benefits and risks. As we have seen, lack of farmer participation may explain why many past interventions failed (Reij 1991; Tiffen et al. 1994; Robbins and Williams 2005). In contrast, participatory interventions incorporating collective action have been relatively more successful (Joshi et al. 2004; Shiferaw et al. 2008b). Technologies resulting from such processes typically take into account the unique socioeconomic characteristics of farmers, allowing adaptation to specific circumstances. Farmers are able to test practices at their own pace and in their preferred sequences, better leading to compatibility with local farming systems (Robbins and Williams 2005). Participatory approaches also allow farmers to gradually adapt technologies to changing conditions (Bunch 1989) and learn from one another.

Markets, Poverty, and Risk

Studies that examine the relationship between commodity prices and land and water management find mixed effects (Barrett 1991; Bulte and van Soest 1999; Lichtenberg 2006). The ambiguous effects are not surprising, because higher commodity prices increase the returns to land management and therefore land value (Lichtenberg 2006), but also can make soil degradation attractive. For instance, increases in the price of agricultural outputs can mask the effect of land degradation, making erosive practices attractive to farmers. When conservation does not provide obvious financial returns, an increase in the price of an erosive crop may encourage expansion without investment in SLM. In other cases, though, increased commodity prices may make SLM profitable for farmers, and a number of studies find positive relationships between prices and adoption (e.g., Bulte and van Soest 1999; Shiferaw and Holden 2000; Lee 2005). Shiferaw and Holden (2000), for example, find that in highland Ethiopia when conservation offers short-term gains, increases in prices spur adoption of SLM.

Government price supports can undermine sustainable land management by distorting incentives faced by resource users. Price supports to irrigated crops like rice and wheat can discourage farmers in semi-arid areas from cultivating sorghum and other water-efficient crops. Well-intentioned policies to promote food security could therefore lead to extensive land degradation and depletion of groundwater resources.

Major determinants of adoption are absolute and relative costs. An increase in the price of fertilizer, for example, generally reduces its application (Pattanayak and Mercer 1997). However, fertilizer subsidies can result in land degradation, as found in China and South Asia (Pingali and Rosegrant 1994; Heerink et al. 2007). Heerink et al. (2007) find, for example, that policies to lower the fertilizer-rice price ratio have lead to compaction and soil degradation. Other studies investigate how the cost of hedgerow cropping, terracing, minimum tillage, no-tillage, and agricultural water harvesting techniques affect adoption and find inverse relationships between cost and adoption (Pattanayak and Mercer 1997; Baidu-Forson 1999; Robbins and Williams 2005). A number of studies examine the role of market access on use of SLM. Most find that when farmers face the costs of land degradation, land rights are clear, and supportive policy and institutional mechanisms exist, improving access to commodity and input markets reduces transaction costs and improves the likelihood of SLM adoption (Reardon et al. 1997; Zaal and Oostendorp 2002). This is perhaps the main conclusion by Hagos and Holden in Chapter 6. They find that, adjusting for a variety of factors, thick and accessible product markets are the key determinants of fertilizer use in the northern Ethiopian regional state of Tigray.

The largely semi-arid Machakos district in Kenya suffered serious soil erosion problems in the 1930s due to failed colonial soil conservation policies, but by the mid-1980s the district had largely brought soil erosion under control while also increasing per capita income (Tiffen et al. 1994; Pagiola 1998; Barbier 2000). This tremendous success has partially been attributed to market access caused by good road infrastructure and proximity to Nairobi (Pagiola 1998; Zaal and Oostendorp 2002; Robbins and Williams 2005). Zaal and Oostendorp (2002) indeed argue that the commercialization of agriculture generated the incomes needed to finance SLM investments. Causality could also be reversed, however. Shiferaw et al. (2008b) find evidence from Adarsha watershed in India that adoption of land management and complementary technologies for improving productivity help farmers diversify into high-value and marketable crops. This suggests that SLM can reduce production risks, increase marketable surplus, and facilitate the transition from subsistence to commercial farming.

The relationship between labor market performance and investments in SLM is quite mixed (Reardon and Vosti 1997; Pender and Kerr 1998; Holden et al. 2004; Robbins and Williams 2005). In the Ethiopian highlands where on-farm returns to family labor are low, Holden et al. (2004) show that increased opportunities for off-farm employment have positive effects on household welfare but reduce conservation investments. Similarly, Shiferaw and Holden (1998) find a negative relationship between off-farm income orientation and maintenance of conservation structures. Pender and Kerr (1998) find that when labor and credit markets work poorly, higher-income households are more likely to invest in SLM. Kerr and Sanghi (1992) find fewer conservation investments around large Indian cities with active off-farm labor markets than in more remote areas. Reardon and Vosti (1997) find similar results in their study of Rwanda, Burundi, and Burkina Faso.

In contrast to these findings, Tiffen et al. (1994), Pagiola (1998), and Scherr (2000) review cases across sub-Saharan Africa where off-farm employment increases soil and water conservation investments, perhaps by reducing the intensity of resource use. But generally the literature finds the opposite, and it offers two main reasons for the negative relationship between labor market performance and SLM investments. First, all else equal, when labor markets work well workers face higher opportunity costs and prefer to allocate labor off-farm. Second, off-farm employment often overlaps with the slack season and reduces labor available for conservation.

Another important factor conditioning adoption and adaptation of conservation technologies is risk. Smallholder farmers face constant difficulties managing health, climate, and socioeconomic shocks; SLM interventions that increase variability or uncertainty of incomes tend to be shunned by farmers. Such risks can arise from greater crop failure (due to biotic and abiotic stresses), poor and unreliable access to markets, or insecure property rights. In Chapter 5, Yesuf and Teklewold find that risk inhibits fertilizer adoption and application. They also find that wealth, far from being an exogenous behavioral factor, is an important determinant of risk aversion when key markets are missing.

Soil and water conservation generally tend to reduce production risks, but there may be circumstances when risks increase. For example, Shiferaw and Holden (1998) find that in Ethiopia soil and stone bunds cause pest infestation and even flooding. An example where SLM reduces risk is water harvesting and irrigation in semi-arid areas used as part of strategies to cope with and adapt to drought and climatic shocks (Shiferaw et al. 2008b). In addition to risks associated with conservation technologies, uninsured production risk may cause farmers to underinvest in all areas, including SLM (de Janvry et al. 1991).

Product, credit, labor, and insurance markets in rural areas of many developing counties tend to be either missing or highly imperfect. Input and output market access is often constrained by poor transport and communication infrastructure and fragmented supply chains, resulting in high transaction costs that undermine commercialization (Fafchamps and Hill 2005; Poulton et al. 2006) and reduced SLM adoption (Pender and Kerr 1998). The importance of market performance for investment is also highlighted in several chapters in this volume.

Using large-scale survey data from Uganda, Pender et al. (2004) test the effect of distance from all-weather roads and markets on commercial crop production and soil erosion. They find that market distance is not correlated with production or erosion. Pender and Kerr (1998) examine the impact on SLM adoption of incomplete and missing input and output markets in semi-arid areas of India. They find that both reduce profitability and adoption.

Access to credit is especially important for adoption of land management interventions like irrigation, terracing, tree planting, and fertilizer use, because of their heavy upfront cash requirements (Holden et al. 1998; Shiferaw and Holden 2000); in most rural areas in East Africa, however, credit markets work very poorly. Households must therefore rely on their own assets, and several studies show that assets (including human capital) influence investments in conservation (Reardon and Vosti 1995; Holden et al. 1998; Scherr 2000; Swinton and Quiroz 2003). The role of education and other forms of human capital on adoption of SLM has been particularly widely studied (e.g., Knowler and Bradshaw 2007). Human capital increases the likelihood that farmers will perceive land degradation as a problem, and it may increase managerial ability, helping farmers process information about technologies. However, if off-farm options like migration and nonagricultural wage employment are available, more education can increase the opportunity cost of labor and reduce incentives to invest (Swinton and Quiroz 2003).

Most land management investments like the *fanya juu* terraces promoted in the Machakos District of Kenya require large initial investments, but they deliver a flow of benefits over many years. Due to imperfect capital markets and associated high costs of borrowing, combined with their own limited resources, most resource-poor farmers have short planning horizons (Holden et al. 1998). These horizons can discourage adoption of technologies that may not offer immediate benefits but improve livelihoods only in the long run (see Figure 2-2).

Using Figure 2-2, let us assume Options 1 through 4 offer different adoption income streams. The resource-degrading practice is Option 1, with incomes falling over time. Under the next best conservation option—Option 2—incomes decline too, but more slowly. As is typical for many land management investments, net income in the first few years is lower than without investment, but higher thereafter. At the same time, such investments tend to generate external benefits that farmers often omit in their computation of investment benefits. For instance, investment in soil conservation can reduce degradation of downstream fishing and irrigation resources. Evidence

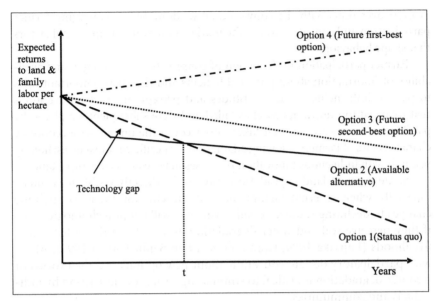

FIGURE 2-2 Challenges in the Design and Development of Pro-poor SLM Technologies

indicates that if farmers face only these two alternatives, resource-conserving technologies are unlikely to be adopted (Holden et al. 1998), because in environments of imperfect markets poor farmers lack the capacity to absorb initial income losses. Unless subsidized, farmers may not be interested in such options (Shiferaw and Holden 2001; Pagiola et al. 2002).

Alternatively, if farmers have access to Options 3 and 4, there will not be such tradeoffs between current and future incomes, and one would expect widespread adoption. A key challenge is that many of the available SLM technologies are not like Options 3 and 4. As particularly emphasized by Menale Kassie in Chapter 8, identifying, developing, and promoting the most suitable SLM technologies and making those approaches incentive-compatible in environments of highly imperfect markets is perhaps the most important challenge facing promoters of SLM.

Conclusions and Policy Implications

This chapter reviews the challenges that small farmers face in tackling land degradation and presents a broad conceptual framework for understanding SLM investments within the context of imperfect factor markets, inadequate property rights, and weak organizational and institutional arrangements. Our review of the literature suggests that resource-poor farmers, especially in marginal and rain-fed regions, face complex challenges in adopting and adapting land management innovations. Approaches to soil and water conservation have evolved over time, with the conventional wisdom now encouraging farmer participation and consideration of the market, policy, and institutional factors that shape behavior.

Farmer participation in the design of conservation technologies and availability of information about potential benefits and risks have important roles to play in influencing farmers' attitudes and perceptions. Past interventions that followed top-down approaches failed and were subsequently replaced by participatory conservation that takes into account the unique socioeconomic characteristics of farmers, allowing adaptation to specific circumstances; linking research with indigenous innovation processes may be especially important.

Some types of land degradation may not be directly visible to farmers, especially when external factors make it difficult for them to attribute changes to declining resource quality. Farmers will adopt technologies only if they perceive soil and water degradation as a problem that affects their livelihoods (Fujisaka 1994; Cramb et al. 1999; Baidu-Forson 1999). Along with participatory design, education about new options and the process of resource degradation is critical to stimulating awareness and action by individuals and communities.

Commercialization of agriculture and better market integration generally raises the returns to agricultural land and labor. When complemented by policies and institutional mechanisms to induce innovation and adoption, thicker, more accessible markets can be important drivers of sustainable intensification. Given that poverty and lack of farmer capacity can be major limiting factors, access to credit at affordable rates and availability of pro-poor, profitable conservation technologies are key steps.

Unless conservation measures provide higher returns and/or lower risks than unsustainable options, farmers cannot be expected to adopt them; several studies show that returns to SLM can be negative (e.g., see Chapter 8 in this volume). In the presence of significant market failures and when the social gains are higher than costs, conservation subsidies may be justified. With pervasive off-site effects and market failures that hinder landscape-wide interventions, stimulating wider use of SLM will also require new kinds of institutional mechanisms for empowering communities through collective action. This chapter has shown that the interests of smallholder farmers and society may not always coincide in attaining social objectives for sustainable use and management of land, water, and other vital resources.

There is a critical need for additional research to identify policies and institutional reforms that overcome market and policy failures in smallholder agriculture and stimulate investments in SLM. One of the most innovative approaches to help poor smallholder farmers adopt more sustainable practices is payment for environmental services (PES). Under PES beneficiaries of environmental services compensate farmers who invest in protection and supply of ecosystem services (Pagiola et al. 2002; Pagiola et al. 2005). Pagiola et al. (2005) find that PES schemes can reduce poverty while internalizing the external benefits of conservation. There is a need to test, develop, and adapt such innovations to create greater incentives for beneficial conservation of land, water, and agro-ecosystems in Africa.

Notes

1   Using studies in Niger, Tabor (1995), for instance, points out that despite holding tremendous promise for increasing crop yields in semi-arid lands, applications of water-harvesting technologies are not terribly widespread.

2   The effect of agricultural price and non-price subsidies and the importance of public investment in transport infrastructure and the associated effects of improved market access and competitiveness on SLM will be discussed in the following section.

References

Ahuja, A. 1998. Land Degradation, Agricultural Productivity, and Common Property: Evidence from Côte d'Ivoire. *Environment and Development* 3: 7–34.

Baidu-Forson, J. 1999. Factors Affecting Adoption of Land-Enhancing Technology in the Sahel: Lessons from a Case Study in Niger. *Agricultural Economics* 20: 231–39.

Barbier, E. 2000. The Economic Linkages between Rural Poverty and Land Degradation: Some Evidence from Africa. *Agriculture, Ecosystem and Environment* 82: 355–70.

Barrett, S. 1991. Optimal Soil Conservation and the Reform of Agricultural Pricing Policies. *Journal of Development Economics* 36: 167–87.

Barrett, C. B., J. Lynam, F. Place, T. Reardon, and A. A. Aboud. 2002. Towards Improved Natural Resource Management in African Agriculture. In *Natural Resource Management in African Agriculture: Understanding and Improving Current Practices*, edited by C. Barrett, F. Place, and A. A. Aboud. Wallingford, UK: CAB Publishing, 287–96.

Biot, Y., P. Blakie, P. Jackson, and R. Palmer-Jones. 1995. Re-Thinking Research on Land Degradation in Developing Countries. Discussion Paper 289. Washington, DC: The World Bank.

Bulte, E., and D. van Soest. 1999. A Note on Soil Depth, Failing Markets, and Agricultural Pricing. *Journal of Development Economics* 58: 245–54.

Bunch, R. 1989. Encouraging Farmers' Experiments. In *Farmer First*, edited by R. Chambers, A. Pacey, and L. Thrupp. London: Intermediate Technology Publications, 55–59.

Chambers, R. 1987. Sustainable Livelihoods, Environment, and Development: Putting Poor People First. Discussion Paper 240. Sussex, UK: Institute of Development Studies.

Chambers, R., A. Pacey, and L. Thrupp, eds. 1989. *Farmer First: Farmer Innovation and Agricultural Research*. London: Intermediate Technology Publications.

Colombo, S., J. Calatrava-Requena, and N. Hanley. 2006. Analyzing the Social Benefits of Soil Conservation Measures Using Stated Preference Methods. *Ecological Economics* 58: 850–61.

Cramb, R. A., J. N. M. Garcia, R. V. Gerrits, and G. C. Saguiguit. 1999. Smallholder Adoption of Soil Conservation Technologies: Evidence from Upland Project from the Philippines. *Land Degradation and Development* 10: 405–23.

de Janvry, A., M. Fafchamps, and E. Sadoulet. 1991. Peasant Household Behaviour with Missing Markets: Some Paradoxes Explained. *Economic Journal* 101: 1400–17.

Fafchamps, M., and R. V. Hill. 2005. Selling at the Farm-Gate or Traveling to Market. *American Journal of Agricultural Economics* 87: 717–34.

Fox, G., G. Umali, and T. Dickinson. 1995. An Economic Analysis of Targeting Soil Conservation With Respect to Off-Site Water Quality. *Canadian Journal of Agricultural Economics* 43: 105–18.

Fujisaka, S. 1994. Learning From Six Reasons Why Farmers Do Not Adopt Innovations Intended to Improve Sustainability of Upland Agriculture. *Agricultural Systems* 46: 409–25.

Gebremedhin, B., J. Pender, and G. Tesfay. 2003. Community Natural Resource Management: The Case of Community Woodlots in Northern Ethiopia. *Environment and Development Economics* 8: 129–48.

Heath, J., and H. P. Binswanger. 1996. Natural Resource Degradation Effects of Poverty and Population Growth are Largely Policy Induced: The Case of Colombia. *Environment and Development Economics* 1: 64–84.

Heerink, N., F. Qu, M. Kuiper, X. Shi, and S. Tan. 2007. Policy Reforms, Rice Production, and Sustainable Land Use in China: A Macro-Micro Analysis. *Agricultural Systems* 94: 784–800.

Hoff, K., A. Braverman, and J. E. Stiglitz, eds. 1993. *The Economics of Rural Organization.* Oxford: Oxford University Press.

Holden, S. T., B. Shiferaw, and M. Wik. 1998. Poverty, Credit Constraints, and Time Preferences: Of Relevance for Environmental Policy? *Environment and Development Economics* 3: 105–30.

Holden, S., B. Shiferaw, and J. Pender. 2004. Non-Farm Income, Household Welfare, and Sustainable Land Management in the Less Favored Area in the Ethiopian Highlands. *Food Policy* 29: 369–92.

Hurni, H. 2000. Assessing Sustainable Land Management (SLM). *Agriculture, Ecosystems and Environment* 81: 83–92.

Joshi, P. K., V. Pangare, B. Shiferaw, S. P. Wani, J. Bouma, and C. Scott. 2004. Watershed Development in India: Synthesis of Past Experiences and Needs for Future Research. *Indian Journal of Agricultural Economics* 59(3): 303–20.

Kelly, V., A. Adesina, and A. Gordon. 2003. Expanding Access to Agricultural Inputs in Africa: A Review of Recent Market Developments. *Agricultural Economics* 28: 379–404.

Kerr, J., and N. Sanghi. 1992. Indigenous Soil and Water Conservation in India's Semi-arid Tropics. Gatekeeper Series No. 34. London: International Institute for Environment and Development.

Kerr, J., G. Milne, V. Chhotray, P. Baumann, and A. J. James. 2007. Managing Watershed Externalities in India. Theory and Practice. *Environment, Development and Sustainability* 9: 263–68.

Knowler, D., and B. Bradshaw. 2007. Farmers' Adoption of Conservation Agriculture: A Review and Synthesis of Recent Research. *Food Policy* 32: 25–48.

Lee, D. R. 2005. Agricultural Sustainability and Technology Adoption in Developing Countries: Issues and Policies for Developing Countries. *American Journal of Agricultural Economics* 87: 1325–34.

Lichtenberg, E. 2006. "A Note on Soil Depth, Failing Markets, and Agricultural Pricing": Comment. *Journal of Development Economics* 81: 236–43.

North, D. C. 1990. *Institutions, Institutional Change, and Economic Performance.* Cambridge: Cambridge University Press.

Pagiola, S., J. Bishop, and N. Landell-Mill. 2002. *Selling Forest Environmental Services: Market-Based Mechanisms for Conservation and Development.* London: Earthscan Publications Limited.

Pagiola, S., A. Arcenas, and G. Platais. 2005. Can Payments For Environmental Services Help Reduce Poverty? An Exploration of the Issues and the Evidence to Date from Latin America. *World Development* 33: 237–53.

Pagiola, S. 1998. Economic Analysis of Incentives for Soil Conservation. In *Using Incentives for Soil Conservation*, edited by D. W. Sanders, P. C. Huszar, S. Sombatpanit, and T. Enters. Enfield, New Hampshire: Science Publishers Inc.

Pandey, S. 2001. Adoption of Soil Conservation Practices in Developing Countries: Policy and Institutional Factors. In *Response To Land Degradation*, edited by E. Bridges, I. Hannam, I. Oldeman, L. Penning De Vries, F. Scherr, and S. Sombatpanit. Enfield, New Hampshire: Science Publishers Inc.

Pattanayak, S., and D. E. Mercer. 1997. Valuing Soil Conservation Benefits of Agroforestry: Contour Hedgerows in Eastern Visayas, Philippines. *Agricultural Economics* 18: 31–46.

Pender, J. L., and J. M. Kerr. 1998. Determinants of Farmers' Indigenous Soil and Water Conservation Investments in Semi-arid India. *Agricultural Economics* 19: 113–25.

Pender, J. L., E. Nkonya, P. Jager, D. Sserunkuuma, and H. Ssali. 2004. Strategies to Increase Agricultural Productivity and Reduce Land Degradation. *Agricultural Economics* 31: 181–95.

Pender, J., and P. Hazell. 2000. Promoting Sustainable Development in Less-Favored Lands: Overview. In *Promoting Sustainable Development in Less-favored Lands*, edited by J. Pender, and P. Hazell. Focus 4. Washington, DC: International Food Policy Research Institute.

Pingali, P. L., and M. W. Rosegrant. 1994. Confronting the Environmental Consequences of the Green Revolution in Asia. EPTD Discussion Paper No. 2. Washington, DC: International Food Policy Research Institute.

Poulton, C., J. Kydd, and A. Doward. 2006. Overcoming Market Constraints on Pro-Poor Agricultural Growth in Sub-Saharan Africa. *Development Policy Review* 24: 243–77.

Reardon, T., and S. A. Vosti. 1995. Links between Rural Poverty and Environment in Developing Countries: Asset Categories and Investment Poverty. *World Development* 23: 1495–1503.

Reardon, T., and S. A. Vosti. 1997. Poverty Environment Links in Rural Areas of Developing Countries. In *Sustainable Growth and Poverty Alleviation*, edited by T. Reardon and S.A. Vosti. Baltimore: The John Hopkins University Press.

Reardon, T., V. Kelly, E. Crawford, B. Diagana, J. Dione, K. Savadogo, and D. Boughton. 1997. Promoting Sustainable Intensification of and Productivity Growth in Sahel Agriculture after Macroeconomic Policy Reform. *Food Policy* 22: 317–27.

Reddy, V. R. 2005. Costs of Resource Depletion Externalities: A Study of Groundwater Overexploitation in Andhra Pradesh, India. *Environment and Development Economics* 10: 533–56.

Reij, C. 1991. Indigenous Soil and Water Conservation in Africa. Gatekeeper Series No. 27. London: International Institute for Environment and Development.

Ribaudo, M.O. 1986. Considerations of Offsite Effects Impacts in Targeting Soil Conservation Programs. *Land Economics* 62: 402–11.

Robbins, M., and T. O. Williams. 2005. Land Management and Its Benefits: The Challenge and the Rationale for Sustainable Management of Dry Lands. Paper presented at A STAP Workshop on Sustainable Land Management. 2005. The World Bank, Washington, DC.

Scherr, S. 2000. A Downward Spiral? Research Evidence on the Relationship between Poverty and Natural Resource Degradation. *Food Policy* 25: 479–98.

Shiferaw, B., and C. Bantilan. 2004. Rural Poverty and Natural Resource Management in Less-Favored Areas: Revisiting Challenges and Conceptual Issues. *Journal of Food, Agriculture and Environment* 2(1): 328–39.

Shiferaw, B., and S. T. Holden. 1998. Resource Degradation and Adoption of Land Conservation Technologies in the Ethiopian Highlands: A Case Study in Andit Tid, North Shewa. *Agricultural Economics* 18(3): 233–48.

———. 2000. Policy Instruments for Sustainable Land Management: The Case of Highland Smallholders in Ethiopia. *Agricultural Economics* 22: 217–32.

———. 2001. Farm-Level Benefits to Investment for Mitigating Land Degradation. *Environment and Development Economics* 6: 355–58.

Shiferaw, B., V. R. Reddy, and S. Wani. 2008a. Watershed Externalities, Shifting Cropping Patterns and Groundwater Depletion in Indian Semi-arid Villages: The Effect of Alternative Water Pricing Policies. *Ecological Economics* 67(2): 327–40.

Shiferaw, B., C. Bantilan, and S. Wani. 2008b. Rethinking Policy and Institutional Imperatives for Integrated Watershed Management: Lessons and Experiences from Semi-arid India. *Journal of Food, Agriculture, and Environment* 6(2): 370–77.

Swinton, S. M., and R. Quiroz. 2003. Is Poverty to Blame for Soil, Pasture, and Forest Degradation in Peru's Altiplano? *World Development* 31(11): 1903–19.

Tabor, J. A. 1995. Improving Crop Yields in Sahel by Means of Water Harvesting. *Journal of Arid Environments* 30: 83–106.

Tiffen, M., M. Martimore, and F. Gichuki. 1994. *More People, Less Erosion: Environmental Recovery in Kenya*. London: John Wiley and Sons Publishers.

Zaal, F., and R. H. Oostendorp. 2002. Explaining the Miracle: Intensification and the Transition to Towards Sustainable Small-Scale Agriculture in Dry Land Machakos and Kitui Districts, Kenya. *World Development* 30: 1271–87.

CHAPTER 3

The Role of Social Capital in Sustainable Development: An Analysis of Soil Conservation in Rural Kenya

WILFRED NYANGENA

In Kenya over 80% of the poor live in rural areas and are dependent on natural resources like soil for their livelihoods. Many parts of Kenya experience severe soil erosion, which contributes to low and declining productivity that can profoundly affect poor households. There are also downstream effects such as water pollution, sedimentation and siltation of water bodies, disruption of aquatic ecology, and destruction of road infrastructure.

Past efforts to address land degradation have focused on changing individual behavior rather than influencing communities. This is surprising, because a focus of the contemporary economic development literature is to identify and understand how social aspects of individuals and communities contribute to or hinder economic performance (Narayan and Pritchett 1999; Stewart 2005); indeed, a growing number of economists acknowledge that social capital can generate streams of future benefits and therefore shares at least some similarities with physical and human capital. Oft cited specific benefits include information sharing, insurance, promotion of cooperation, and trust that increases efficiency and collective action (La Porta et al. 1997).[1]

Recent studies also try to identify the determinants of social capital formation (e.g., La Ferrara 2004). In this chapter, we use common indicators to extend the literature and complement several chapters in this volume by investigating the factors associated with social capital in three districts in Kenya. We then extend traditional models of technology adoption to investigate the link between social capital and adoption of soil and water conservation (SWC) technologies.

The remainder of the chapter proceeds as follows. We discuss the reasons social capital may matter for soil conservation in the next section. We then describe the data and social capital variables. Next we present the estimation

31

strategy and discuss results in the following section. The final section concludes by drawing some key policy implications.

Literature

Much about the determinants of SWC adoption in developing countries remains unclear. Early studies focused on individual and plot characteristics (e.g., Feder et al. 1985), but more recent theoretical work emphasizes the importance of social arrangements, incentive structures and growth (e.g., Fershtman et al. 1996; Foster and Rosenzweig 1995; Rogers 1995). There are several classes of models explaining soil management, and many are dynamic and emphasize the role of soil as capital. These models link and investigate the effects of economic factors like market imperfections and price incentives, as well as biophysical factors like soil depth, on soil capital (see McConnell 1983; LaFrance 1992; Goetz 1997; Yesuf 2004). While these studies show that economic and biophysical factors are very important, the links between economic factors and soil conservation outcomes remain important areas of research.

It is also well-known that social capital can foster cooperative behavior and reduce coordination problems (Krishna 2001; Bowles and Gintis 2002). Structures of social relations may enable trust, which can allow people to coordinate their actions for mutual benefit (Ostrom 1990). Pretty (2003), for example, documented the importance of social capital for a range of natural resource management sectors, including watershed management, integrated pest management, and farmer experimentation.

The notion of social capital came to the fore with the much-publicized work of Putnam (1993), and since that time a small but well-established literature on developing countries has emerged. Many emphasize the impact of social capital on outcomes like economic growth (Knack and Keefer 1997), incomes (Narayan and Pritchett 1999), and greater use of modern agricultural inputs (Isham 2002). Social capital also serves to mitigate effects of individual-specific economic shocks (Carter and Maluccio 2003; Fafchamps and Lund 2003; Krishna 2001; Krishna and Uphoff 1999).

As several chapters in this volume note, construction of soil conservation structures is demanding, and local farmers may be poorly equipped if they have little formal training or poor access to good agricultural extension services. Group coordination may therefore help exploit economies of scale and scope associated with soil conservation, including maintaining links with government agencies. Construction of SWC structures also demands a lot of labor, which can make them unattractive for many households. Farmers therefore may rely on labor pooling to overcome labor shortages.

Formal credit markets do not function well in many agricultural societies due to high information, monitoring, and transaction costs, lack of collateral, and moral hazard problems (Stiglitz and Weiss 1981). This lack of credit may discourage investments, but strong social capital can facilitate the pooling of finances, which can then be invested in soil conservation. Benefits from soil conservation also materialize with a lag. With no possibility to save or borrow, investments are made at the expense of current consumption (Hoff et al. 1993). Under these circumstances social ties can help with consumption smoothing. Rural farmers also operate under imperfect information with limited awareness of markets and technologies. One practical aspect of social capital is the ability to provide information channels on markets that may be relevant for SWC investment decisions.

Finally, while SWC technologies are employed on individual farms, as emphasized in Chapter 10 by Anders Ekbom and in Chapter 2 by Bekele Shiferaw and Julius Okello, positive externalities from conservation make collective action important. For instance, as soil erosion does not respect boundaries, farm technologies like terracing require widespread and coordinated adoption in order to be effective. Without coordination, even farmers who adopt SWC measures may still face uncontrollable damage from neighboring farms.

The products of SWC investments are therefore a complex mixture of public, private, and club goods with associated incentives to free ride. These investments are also made within environments where key markets work poorly. Social institutions based on trust, reciprocity, and rules can substitute for markets and mediate negative effects of unfettered private action.

Data

The data were collected in the Machakos, Kiambu and Meru districts from January through April of 2003. From these we randomly selected 10 villages and 20 households from each village to be surveyed. Questions are based on World Bank studies of social capital, poverty, and development (see www.worldbank. org/poverty/scapital/index.htm). Surveys include questions on demographics, human capital, land under cultivation, assets, access to markets and infrastructure, community variables and plot level[2] agricultural practices (e.g., crops and acreage, output, prices, SWC types) for the 2001/2002 production season. We also collect information about relationships, memberships in voluntary groups and associations, monetary and in-kind contributions, and sources of private and public information. The questionnaire used was pre-tested and administered to household heads. It is included in appendix 3.1 at the end of this chapter.

A number of studies on soil conservation employ dichotomous variables to represent adoption decisions (Feder et al. 1985; Place and Hazell 1993; Shiferaw and Holden 1998), including Chapter 4 of this book by Kabubo-Mariara and Linderhof. We also use a dichotomous variable (*Conserve*) that takes a value of one if a physical conservation structure was present on a plot during the last five years and zero otherwise. Structures include bench terraces, *fanya juu*, and infiltration ditches. We find that the proportion of plots with

Wilfred Nyangena

TABLE 3-1 Descriptive Statistics of Variables

Variable	Definition	Kiambu		Machakos		Meru		All	
Dependent variable		Mean	Std.dev	Mean	Std.dev	Mean	Std.dev	Mean	Std.dev
CONSERVE	Presence of SWC structure on plot	0.53	0.64	0.67	0.44	0.64	0.43	0.57	0.49
Farm Characteristics									
HIGH EROSION	Proportion of highly eroded plots	0.23	0.42	0.06	0.24	0.19	0.36	0.19	0.39
MODERATE EROSION	Proportion of mildly eroded plots	0.04	0.09	0.25	0.42	0.13	0.09	0.12	0.11
LOW EROSION	Proportion of least eroded plots	0.74	0.44	0.65	0.48	0.67	0.41	0.69	0.34
UPPER SLOPE	Plot located in upper slope = 1, Else=0	0.39	0.48	0.37	0.47	0.35	0.48	0.37	0.48
MID SLOPE	Plot is located in mid slope = 1, Else=0	0.31	0.46	0.29	0.46	0.48	0.50	0.35	0.47
LOW SLOPE	Plot is located in lower slope = 1, Else=0	0.29	0.45	0.33	0.49	0.16	0.37	0.27	0.41
Tenure Security									
HIGH	Complete rights = 1, Else = 0	0.71	0.45	0.12	0.33	0.66	0.47	0.51	0.68
MEDIUM	Preferential use rights = 1, Else = 0	0.16	0.37	0.62	0.48	0.15	0.35	0.21	0.41
LOW	Limited use rights = 1, Else = 0	0.13	0.34	0.25	0.44	0.19	0.39	0.15	0.36
Behavioral/Household Characteristics									
EDUCATION	Years of schooling for all above 16	7.6	2.56	6.14	2.28	6.77	2.27	7.16	2.48
AGE HH	Age of the household head in years.	52.8	14.3	55.13	10.61	47.91	12.51	51.9	13.5
DEPENDENCY RATIO	Ratio adults to < 6 and > 65 in family	0.32	0.20	0.31	0.22	0.28	0.20	0.31	0.21
HIRED LABOR	Share hired farm workers	0.31	0.57	0.08	1.29	0.06	1.43	3.2	1.5
REMITTANCES	Receipt of remittances = 1, Else = 0	0.69	0.46	0.90	0.30	0.61	0.95	0.77	0.42
PER CAPITA LAND	Share of land area to family size	0.30	0.32	0.29	0.27	0.26	0.18	0.28	0.27
PERENNIAL CROP	Perennial crop on plot = 1, Else = 0	0.28	0.45	0.31	0.46	0.41	0.49	0.32	0.47

	0.31	0.18	0.69	0.25	0.51	0.36	0.51	0.23
PRIOR ADOPTION								
Distance to								
Proportion of previous adoption (%)								
PRODUCE MARKET								
Mean walk time to nearest market (Min)	33.00	16.00	60.00	55.00	31.00	18.00	35.0	24.0
ADMINISTRATIVE HQ								
Bus fare home to divisional centre (Kshs)	56.00	9.00	47.00	8.00	42.00	27.00	47.0	21.0
Sample Size (number of plots)								
ADOPTERS	183		44		93		320	
NONADOPTERS	162		22		52		236	

SWC is highest in Machakos and lowest in Kiambu. Previous studies have also found that adoption is highest in Machakos (Tiffen et al. 1994). Descriptive statistics are presented in Table 3-1.

The choice of covariates in the model is based on a review of the determinants of adoption, which others have found to be significant (Ervin and Ervin 1982; Feder et al. 1985; Besley and Case 1993; Shiferaw and Holden 1998; Lapar and Pandey 1999; Gebremedhin and Swinton 2003). Topographic and farm characteristics are seen as especially important. The proportion of highly eroded plots in Kiambu and Meru is similar at about 20%, but Machakos has only 6% highly eroded plots. The position of a plot on the slope profile, also known as its catena, is an important indicator of erosion potential and soil conditions (Lapar and Pandey 1999). On a typical slope the steepest region is found mid-slope. Thus, one would expect short-run productivity loss to be highest, all else equal, in the middle catena. Hence, plots on the mid-slope catena might have more conservation investments. The proportion of plots located in the upper slope is comparable across regions at 35% to 37%, but there is a remarkable difference in the proportion of plots located mid-slope, with Meru highest and Machakos lowest.

Household characteristics may also affect conservation investment decisions, because of the substantial labor costs and imperfect markets. To capture the effects of age composition we include a dependency ratio that is equal to the number of persons who cannot work (under age 6 and above 65) divided by total household members. Other household characteristics include years of education, age (linear and quadratic terms), and gender of household heads. On average the youngest household head is found in Meru. The mean number of years of formal education is lowest in Machakos (6.14 years) and highest in Kiambu (7.6 years).

Adoption is a gradual process that involves sequential stages. Economists have attempted to develop models for evaluating adoption and lag between the initial awareness and the actual use by farmers. In this study we use prior adoption to capture the lag between past and current adoption. The proportion of households with previous adoption is highest in Machakos (69%), followed by Meru (51%) and Kiambu (31%).

Land holding per capita is low, which is similar to other high density regions of Kenya, and almost uniform across districts. The proportion of households receiving remittances is highest in Machakos (90%) and lowest in Meru (61%). Poor land tenure has sometimes been blamed for low SWC investments, which would suggest that tenure in Machakos should be most secure. In fact, we find that Machakos cannot readily be described as high tenure security. Access to markets also differs. Farmers in Machakos walk one hour to sell their produce at a regional market, while travel time from Meru and Kiambu is about 30 minutes.

Though social capital is recognized as being an important element of resource management, it remains a difficult issue to address empirically

(Paldam and Svendsen 2004). We sought social capital indicators to measure (1) interactions based on borrowing small farm implements, mutual risk coping strategies, and SWC information sharing; (2) working with neighbors and participation in local collective action, and (3) sources of market and public information.

Most of the social capital literature specifically mentions trust as an important element. Unfortunately, we could not ask respondents the extent to which they trust neighbors. Presumably, however, activities like borrowing money or food from nonrelatives would be more common in environments where people trust and we therefore use this as a proxy.

Studies by Krishna (2001) and Narayan and Pritchett (1999) ask questions[3] about households' social relations, memberships in groups, participation in community activities, and attitudes and values in settings similar to ours. Responses from these questions are then combined to form quantitative indicators of social capital using factor analysis, a method of data reduction that describes indicators as linear combinations of a small set of underlying variables (Dunteman 1994; Johnson and Wichern 2002).

We adopt a similar approach, using principal component analysis (PCA), because we have no *a priori* theoretical basis for choosing particular measures of social capital. This is a statistical method that can statistically (not subjectively) identify and weight indicators to calculate aggregate indices of social capital. We use PCA to isolate and measure the social capital component embedded in various variables and create household-specific social capital scores. As a first step we compute Pearson correlation coefficients of our 35 social capital variables and drop weakly correlated variables from the analysis. Table 3-2 presents the factor loadings.

It turns out that many variables are highly collinear and significant. The analysis also uncovers patterns and associations by looking at loadings on variables across components of the variation. Loadings are then used as weights, yielding an overall social capital measure that is a sum of the product of component scores.[4] In the PC analysis a component is retained as long as its eigenvalue is greater than one. In our analysis four components are retained.

To interpret the weights we set the minimum criterion for acceptance at 0.3. As the data are collected at the household level, we compute indices at that level and then average the values to obtain a regional average index. We then create association, solidarity, community, and information measures of social capital. The largest amount of variation is captured by the degree of participation in local associations and community projects, all indicators of strong connections with neighbors. The solidarity index reflects cooperation in the form of lending money, food, and reciprocity to reduce the effects of adverse shocks. These first two indices approximate Putnam's (1993) components of trust and civic engagement (Glaeser et al. 2002).

The third loading component focuses on sharing farm tools and assisting neighbors. It captures volunteerism and goes beyond a shared sense of

TABLE 3-2 Loadings on the First Four Principal Components

Variable		Factor1	Factor2	Factor3	Factor4
Membership (yes/no)	C1	0.363	−0.119	0.195	−0.061
Number of associations	C2	0.508	−0.163	0.161	−0.034
Number of meetings	C3	0.421	−0.159	0.053	−0.062
Monetary contribution to Assoc.	C4	0.236	−0.037	0.081	0.169
Benefits received	C5	0.272	−0.159	−0.038	0.176
Number of close friends	T1	0.155	0.345	−0.196	0.115
# of persons to help in econ. crisis	T2	0.221	0.513	−0.206	−0.079
# of persons to help with crop loss	T3	0.274	0.431	−0.245	−0.099
Value of assistance given last year	T4	0.015	0.088	0.001	−0.143
Lent tools to neighbors	N1	−0.049	0.347	0.565	−0.058
Borrowed tools from neighbors	N2	−0.076	0.302	0.588	−0.074
Prepared to contribute time	N3	−0.085	0.038	−0.251	0.103
Prepared to contribute money	N4	−0.029	−0.099	0.028	0.062
Participated in community project	N5	0.339	0.019	0.069	−0.029
Main source of market info: Media	I1	0.016	0.121	−0.107	0.124
Main source of info: Relatives	I2	0.101	−0.069	0.046	0.505
Main source of info: Commune	I3	0.034	−0.211	−0.010	−0.475
Main source of Gov. info: Relatives	I4	0.104	0.082	−0.104	0.035
Main source of Gov. info: Media	I5	−0.019	−0.087	0.155	0.447
Main source of Gov. info: Public	I6	0.033	−0.172	−0.043	0.399

community (Paxton 1999). Lastly, a component reflecting how farmers collect information is formed by counting and ranking households' most important sources of information on crop prices, agricultural news, and government news. Households with all three sources of information are ranked first, and a household with only one is ranked last. A high value of this variable reflects someone who has many different sources of information and is well connected. Table 3-3 presents the descriptive statistics at the district level, which are used in a multivariate analysis to understand how social aspects of individual and community behavior contribute to or detract from SWC investments.

Some big differences between Machakos and the other regions are apparent. While the mean "solidarity" index is similar, "association" in Machakos is 43% higher than Kiambu and 37% greater than Meru. The maximum value of "solidarity" is twice as high in Machakos as Kiambu. Interestingly, Machakos ranks lowest with regard to "community" and "information"[5] indices.

TABLE 3-3 Descriptive Statistics of Social Capital Indices

Variable	Kiambu Mean	Kiambu Min–Max	Machakos Mean	Machakos Min–Max	Meru Mean	Meru Min–Max	All Mean	All Min–Max
Association	2.34	0 12.6	4.11	0 16	2.60	0 8	3.02	0 16
Solidarity	4.26	2 9	4.49	1 18	4.31	1.6 11	4.36	2 18
Community	0.96	0 1.1	0.92	0 1.4	1.08	0 1.2	0.99	0 1.4
Information	3.34	–4 13.5	2.96	–2.3 10.3	3.01	–3 10.2	3.11	–4 13.6

Modelling and Estimation Issues

Empirical models describing farm technology adoption are typically based on the assumption that households choose to adopt when the present value of future net returns is greater than returns from non-adoption. Social capital affects these returns by shaping opportunities and constraints for farmers and mitigating market distortions that increase costs and reduce the profitability of agricultural production (Zak and Knack 2001). Individuals and regions endowed with social capital choose to engage in collective action and coordination when it increases opportunities and reduces costs.

To highlight these points formally, let h denote the household and p the number of plots. The household makes a decision to invest in SWC on a plot as a function of observable and unobservable household characteristics as shown in (3.1).

$$Y_{hp}^* = \beta'X_{hp} + \varepsilon_{hp} \quad Y_{hp} = 1\,(if\ Y_{hp}^* > 0), \tag{3.1}$$

where Y_{hp} is an observed binary (latent) variable indicating a household's decision to invest in SWC. The vector X_{hp} includes explanatory variables for observable household characteristics that influence the decision to invest. Lastly, β is a vector of coefficients to be estimated, and ε_{hp} is the error term, which is assumed to be random.

The data set consists of single and multiple plots managed by households. There is therefore potential for correlation among plot observations, reducing standard errors and biasing estimated coefficients. A method that accounts for such single and multiple plot-level data is the random effects probit, which supposes household-specific, but plot invariant, characteristics (Wooldridge 2002). In addition, random effects probit models can be used to analyze data that include single plot observations, which is an issue for us (Greene 2000).

The random effects probit model assumes that the correlation between disturbances for individual plots can be reduced to a constant ρ (Butler and Moffit 1982). The relationship in (3.1) is therefore modified to account for multiple plots as:

$$y_{hp}^* = \beta' X_{hp} + v_h + \mu_{hp},$$

$$\varepsilon_{hp} = v_h + \mu_{hp} \text{ and } \text{var}\left[v_h + \mu_{hp} \right] = \text{var}\left[\varepsilon_{hp} \right] = \sigma_v^2 + \sigma_\mu^2. \qquad (3.2)$$

where the set of unobservable characteristics v_h are household-specific attributes that influence farm investment decisions. The correlation between two successive error terms for plots belonging to the same household is a constant estimated as (3.3).

$$\text{corr}\left[\varepsilon_{hp}, \varepsilon_{hp-1} \right] = \rho = \sigma_\mu^2 \big/ \left(\sigma_\mu^2 + \sigma_v^2 \right). \qquad (3.3)$$

The estimated correlation across plots is evaluated using a simple t-test (Greene 1995). If the data are not consistent with the random effects model, the estimate of ρ will turn out to be negligible. Of special concern is substantial variation in plot characteristics held by households. This may generate correlation across plots, which may deflate standard errors and bias the estimated coefficients. We find that tests justify the use of the random effects probit and are also consistent as evidenced in a linear probability model.

Our data cover soil conservation at various dates in the past, whereas social and economic variables are current, which may lead to biased estimates due to changing farmer characteristics. Besley and Case (1993) and Besley (1995) are perhaps the best examples of dealing with this problem. Following Besley (1995) we use investments undertaken during the last five years. Although explanatory variables can change, it is highly unlikely that they would change dramatically during five years. Identification is also facilitated, because our data are at the plot level. As it is unlikely that adoption is in equilibrium, we solve the problem by including prior adoption (earlier SWC investments) for each plot.

For individual farmers social capital is a private asset they can draw on, but there are also public good aspects, because individuals benefit indirectly by living in societies with ample social capital. This creates an issue known as the reflection problem (Manski 2000), because farmer and village behavior are simultaneously determined (Durlauf 2002). Following Manski (2000) and Durlauf and Fafchamps (2004) we tackle the reflection problem by including a lag in the transmission of social effects and excluding the individual from the village average.

Results and Discussion

Using (3.2), Table 3-4 reports marginal effects of explanatory variables on the probability of SWC adoption evaluated at sample means. For comparison, we also present probit results with clustering to correct standard errors due to possible interdependence of plots. A likelihood ratio test of the null hypothesis that the ρ coefficient is zero yields a χ^2 value of 63.5, which is significant at

TABLE 3-4 Estimated Coefficients of Probit and Random-Effects Probit Models of SWC Investment

Variable	Probit		Random-Effects Probit	
	Estimated Coefficient	Marginal Effects	Estimated Coefficient	Marginal Effects
Social capital characteristics				
Individual level social capital				
ASSOCIATIONS	0.174	0.022	0.219**	0.024
SOLIDARITY	0.193*	0.023	0.187*	0.036
COMMUNITY	0.236**	0.034	0.228	0.021
INFORMATION	−0.341	−0.021	−0.377	0.035
Physical and farm characteristics				
Soil erosion status				
(ref. LOW EROSION)				
HIGH EROSION	1.187	0.428	0.587	0.222
MODERATE EROSION	1.509	0.319	1.942	0.527
Location on toposequence				
(ref. LOWER)				
UPPER SLOPE‡	0.078	0.065	0.044*	0.071
MID-SLOPE‡	0.210	0.046	0.237	0.056
Perceived tenure security (ref. HIGH)				
MEDIUM	0.031	0.066	−1.019	−0.078
LOW‡	−1.193	0.125	−1.124**	−0.118
Socioeconomic characteristics				
EDUCATION	−0.175**	−0.031	−0.154**	−0.099
AGE HOUSEHOLD HEAD	0.134*	0.015	0.179*	0.022
AGE SQUARED	−0.012	−0.002	−0.013	−0.002
DEPENDENCY RATIO	0.354	0.172	0.734*	0.136
HIRED LABOR	0.262	0.069	0.322	0.002
REMITTANCES	−0.523*	−0.108	−0.984**	0.130
PER CAPITA LAND	−1.618**	−0.274	−1.669**	−0.163
PERENNIAL TREE CROPS‡	−1.145*	−0.212	−1.276**	−0.132
DISTANCE TO MARKET	−0.013**	−0.002	−0.017**	−0.028
BUS FARE TO DIVISION	−0.002	−0.005	−0.041	−0.014
PRIOR ADOPTION	−0.423	−0.198	−0.481*	−0.215
District dummies†				
MACHAKOS (1 = YES, 0 = NO)	0.973*		0.988	
MERU (1 = YES, 0 = NO	0.765		1.121	
Constant	−1.485*		−1.345*	
Regression diagnostics	−	−	0.927	
Rho				
Log–Likelihood	−238.97		−237.31	
Wald Chi square (25)	216.32		95.34	
# of observations	556		556	

Legend: Partial derivatives are in probability units.
** and * significant at the 1% and 5% level.
‡ For dummy variables marginal effect is a discrete change from 0 to 1.
†Default district is KIAMBU.

much greater than the 1% level. This result suggests the plot variance component is not negligible and consequently the random effects model is justified.

Our primary variables of interest are *Associations, Solidarity, Community,*
and *Information.* The results of both models suggest that some measures of
social capital enhance the likelihood of investing in soil conservation. We find
that in the random effects model *Solidarity* and *Associations* are positively
and significantly correlated with SWC adoption. In the probit, *Associations*
and *Community* are significant. Marginal effects are between 2.3% and
3.6%. *Associations* describe group membership in voluntary organizations.
Our positive relationship with SWC investment may indicate membership
in cooperative societies, which are economically oriented and thus provide
technical assistance and credit. For example, cooperatives pool resources to
improve commercialization of agricultural produce, act as guarantors of
informal loans through rotating credit schemes, exchange farm implements
and information, and are sometimes the primary means through which extension services operate. They also provide a ready source of pooled labor and
in some cases credit under reciprocal arrangements. Other studies also find
membership in local networks to be positively correlated with SWC adoption
(Gabunada and Barker 1995; Swinton and Quiroz 2003).

With respect to *Solidarity* the results suggest that people rely on friends and
other community members to pool risk, which is critical for soil conservation
investment. There is therefore an assurance of consumption smoothing. Such
insurance is important, because farmers undertake investments in which they
may have limited experience. Finally, with higher levels of *Solidarity* there are
fewer coordination problems and related costs across farms, which reduces
spatial externalities. *Information* is not statistically significant at standard levels.

We include district dummies for Machakos and Meru to control for
regional differences. These coefficients are positive, but only in the probit
model is Machakos significant, which suggests that the random effects
approach does a good job of capturing unobservables. Including social factors
is also one way to identify underlying factors, and several studies suggest the
use of social interactions at higher aggregation levels (e.g., Place et al. 2002).

We find that low security is negatively correlated with conservation investment decisions, which likely occurs because tenure security gives the assurance
of retaining the long-term gains from land enhancing investments. This result is
consistent with a number of other chapters in this volume as well as the literature
on tenure security and SWC improvements (Besley 1995; Shiferaw and Holden
1998; Gebremedhin and Swinton 2003). Plot location is also a significant determinant of SWC adoption. Steeper slopes are more vulnerable to erosion and
landslides, suggesting that the marginal benefits of SWC may be higher. We find
the estimated coefficient for *Upper Slope* is positive and indicates that the greater
need for SWC on steeper slopes dominates unobserved factors.

Farmers with larger land holdings are less likely to invest in soil conservation and those with less area per capita are more likely to invest. A number

of explanations are possible. Maintaining per capita food production may induce intensification, increasing SWC investments. Greater land scarcity may also encourage careful management and imply more labor for construction of physical structures. By contrast, in the Philippines it is found that small farm size is a barrier to land conservation investments (Shivley 1999). *Tree crops* discourage soil conservation investment as expected. Tree crops provide soil cover, substituting for SWC structures (Young 1997). Soule and Shepherd (2000) report similar findings.

Other significant variables include *Distance to Markets*, which is an indicator of market transaction costs. We find that increased costs reduce farm profitability and inhibit soil conservation investments; as discussed in several other chapters, reducing transport costs therefore seems to promote adoption of land management practices in rural areas (Binswanger and McIntire 1987; Pender et al. 2004).

There is a concave relationship between household head age and investments in SWC, possibly because younger farmers are stronger and better able to devote labor to SWC investments and have longer planning horizons. This result is consistent with others in the literature (Lapar and Pandey 1999; Shiferaw and Holden 1998). Households with *Remittances* are less likely to adopt soil conservation measures, possibly because extra earning opportunities reduce the time for farm work or relax liquidity constraints (World Bank 1994). Additionally, such households may have little concern about land quality due to focuses on off-farm activities.

The estimated coefficient on *Prior Adoption* is negative and significant. This result is consistent with the hypothesis that farmers learn from others and share unobserved determinants of adoption or economies of scale in input supply. Similar results have been reported for the case of crossbred-cow technology in Tanzania (Abdulai and Huffman 2005).

Conclusion

Many natural resource management, agriculture, and marketing projects, programs, and policies are supported by governments and development partners, but in some cases these efforts have met with significant resistance and limited success. Promotion of sustainable land and water management in East Africa is certainly no exception. Our results suggest that to avoid the failures of past projects it is important to subject these policies to rigorous tests of social arrangements. Several dimensions of social capital are likely important for SWC investment, suggesting that social capital should form part of public policies to upscale adoption of SWC technologies. These findings are consistent with recent evidence that household economic performance and collective action are increasing in social capital (Narayan and Pritchett 1999; Krishna and Uphoff 1999; Krishna 2001; Carter and Maluccio 2003).

We also identify other factors that play roles in SWC adoption. SWC adoption varies with farm and household characteristics, suggesting the need for targeted promotion. For example, government interventions promoting farm technology should deliberately target younger farmers. We also find, as have others, that tenure security is important for SWC investments. With better security there are incentives to build terraces, because farmers can enjoy benefits over a long period of time. Though we find little evidence of an impact of access to administrative centers, access to markets is critical for SWC adoption. These results imply that improving infrastructure to reduce transportation costs spurs SWC investments.

Acknowledgment

Financial support from the Swedish International Development Cooperation Agency (Sida) is highly appreciated.

Notes

1   This set of ideas is not without skeptics (e.g., Portes 1998; Durlauf 2002; Dasgupta 2005).
2   A plot is a piece of land that has been cultivated with a specific crop or crop combination for which the farmer can measure the inputs and outputs.
3   See World Bank Social Capital Initiative: http://tinyurl.com/overview-social-capital.
4   If X1, X2 ... Xn are the original set of n variables, then a variable Y formed from a linear combination of these takes the form $Y = a1X1 + a2X2 + ... + anXn$ where the ai's (i = 1,2 ... n) are the principal component loadings or weights. The weights or loadings add up to one.
5   The information index has a negative minimum due to the negative weight attached to communal sources of information, presumably seen as substitutes for other sources of information.

References

Abdulai, A., and W. E. Huffman. 2005. The Diffusion on New Agricultural Technologies: The Case of Crossbred-Cow Technology in Tanzania. *American Journal of Agricultural Economics* 87(3): 645–59.
Besley, T., and A. Case. 1993. Modelling Technology Adoption in Developing Countries. *American Economic Review* 83(2): 396–402.
Besley, T. 1995. Property Rights and Investment Incentives: Theory and Evidence from Ghana. *Journal of Political Economy* 103(5): 903–37.
Binswanger, H. P., and J. McIntire. 1987. Behavioral and Material Determinants of Production Relations in Land Abundant Tropical Agriculture. *Economic Development and Cultural Change* 36(1): 73–99.

Bowles, S., and H. Gintis. 2002. Social Capital and Community Governance. *The Economic Journal* 112: 419–36.

Butler, J.S., and R. A. Moffit. 1982. A Computationally Efficient Quadrature Procedure for the One Factor Multinomial Probit Model. *Econometrica* 50: 761–64.

Carter, M., and J. Maluccio. 2003. Social Capital and Coping with Economic Shock: An Analysis of Stunting of South African Children. *World Development* 31(7): 1147–63.

Dasgupta, P. 2005. Economics of Social Capital. *Economic Record* 81(1): 2–21.

Durlauf, S. 2002. On the Empirics of Social Capital. *Economic Journal* 112: 459–79.

Durlauf, S., and M. Fafchamps. 2004. Social Capital. NBER Working Paper No. 10485. www.nber.org/papers/w10485 (accessed September 16, 2007).

Dunteman, G. H. 1994. *Principal Components Analysis.* New Delhi: Sage Publications.

Ervin, C. A., and D. E. Ervin. 1982. Factors Affecting the Use of Soil Conservation Practices: Hypothesis, Evidence, and Empirical Implications. *Land Economics* 58(3): 277–93.

Fafchamps, M., and S. Lund. 2003. Risk Sharing Networks in Rural Philippines. *Journal of Development Economics* 24(3): 427–48.

Feder, G., R. E. Just, and D. Zilberman. 1985. Adoption of Agricultural Innovations in Developing Countries: A Survey. *Economic Development and Cultural Change* 33(2): 255–98.

Fershtman, C., K. M. Murphy, and Y. Weiss. 1996. Social Status, Education, and Growth. *Journal of Political Economy* 104: 108–32.

Foster, A., and M. Rosenzweig. 1995. Learning by Doing and Learning from Others: Human Capital and Technical Change in Agriculture. *Journal of Political Economy* 103(6): 1176–1209.

Gabunada, F., and R. Barker. 1995. Adoption of Hedgerow Technology in Matalom, Leyte, Philippines. Mimeo.

Gebremedhin, B., and S. Swinton. 2003. Investment in Soil Conservation in Northern Ethiopia: The Role of Land Tenure Security and Public Programs. *Agricultural Economics* 29: 69–84.

Glaeser, E. L., D. Laibson, and B. Sacerdot. 2002. An Economic Approach to Social Capital. *Economic Journal* 112: 431–58.

Goetz, R. U. 1997. Diversification in Agricultural Production: A Dynamic Model of Optimal Cropping to Manage Soil Erosion. *American Journal of Agricultural Economics* 79: 341–56.

Greene, W. H. 1995. *LIMDEP Version 7.0 Users Manual.* BellPort, NY: Prentice Hall.

Greene, W. H. 2000. *Econometric Analysis.* Fourth Edition. Upper Saddle River, NJ: Prentice Hall.

Hoff, K., A. Braverman, and J. Stigliz. 1993. *The Economics of Rural Organization: Theory, Practice, and Policy.* New York: Oxford University Press.

Isham, J. 2002. The Effect of Social Capital on Fertiliser Adoption: Evidence from Rural Tanzania. *Journal of African Economies* 11(1): 39–60.

Johnson, R. A., and D. W. Wichern. 2002. *Applied Multivariate Statistical Analysis.* Prentice Hall.

Knack, S., and P. Keefer. 1997. Does Social Capital Have a Pay-off? A Cross-Country Investigation. *Quarterly Journal of Economics* 112: 1252–88.

Krishna, A. 2001. Moving from the Stock of Social Capital to the Flow of Benefits: The Role of Agency. *World Development* 29: 925–43.

Krishna, A., and N. Uphoff. 1999. Mapping and Measuring Social Capital: A Conceptual and Empirical Study of Collective Action for Conserving and Developing Watersheds in Rajasthan, India. Social Capital Initiative Working Paper No. 13. Washington DC: The World Bank.

LaFrance, J. T. 1992. Do Increased Commodity Prices Lead to More or Less Soil Degradation? *Australian Journal of Agricultural Economics* 36(1): 57–82.

La Ferrara, E. 2004. Kin Groups and Reciprocity: A Model of Credit Transactions in Ghana. *American Economic Review* 93(5): 1730–51.

Lapar, L. A. M., and S. Pandey. 1999. Adoption of Soil Conservation: The Case of the Philippine Uplands. *Agricultural Economics* 21: 241–56.

La Porta, R., F. A. Lopez-de Silanes, and R. W. Vishny. 1997. Trust in Large Organizations. *American Economic Review* 87(2): 333–38.

Manski, C. 2000. Economic Analysis of Social Interactions. *Journal of Economic Perspectives* 14(3): 115–36.

McConnell, K. E. 1983. An Economic Model of Soil Conservation. *American Journal of Agricultural Economics* 65: 83–89.

Narayan, D., and L. Pritchett. 1999. Cents and Sociability: Household Income and Social Capital in Rural Tanzania. *Economic Development and Cultural Change* 47(4): 871–97.

Ostrom, E. 1990. *Governing the Commons: The Evolution of Institutions for Collective Action*. New York: Cambridge University Press.

Paldam, M., and G. T. Svendsen, eds. 2004. *Trust, Social Capital, and Economic Growth: An International Comparison*. Cheltenham: Edward Elgar.

Paxton, P. 1999. Is Social Capital Declining? A Multiple Indicator Assessment. *American Journal of Sociology* 105: 88–127.

Pender, J., P. Jagger, E. Nkonya, and D. Sseunkuuma. 2004. Development Pathways and Land Management in Uganda. *World Development* 32(5): 767–92.

Place, F., and P. Hazell. 1993. Productivity Effects of Indigenous Land Tenure Systems in Africa. *American Journal of Agricultural Economics* 75: 10–19.

Place, F., B. M. Swallow, J. Wangila, and C. B. Barrett. 2002. Lessons for Natural Resource Management Technology Adoption and Research. In *Natural Resources Management in African Agriculture*, edited by C. B. Barrett, F. Place, and A. A. Aboud. Wallingford, UK: CAB International.

Portes, A. 1998. Social Capital: Its Origins and Applications in Modern Sociology. *Annual Review of Sociology* 24: 1–24.

Pretty, J. 2003. Social Capital and the Collective Management of Resources. *Science* 302: 1912–14.

Putnam, R. 1993. *Making Democracy Work: Civic Traditions in Modern Italy*. Princeton, NJ: Princeton University Press.

Rogers, E. M. 1995. *Diffusion of Innovations*. New York: The Free Press.

Shiferaw, B., and S. Holden. 1998. Resource Degradation and Adoption of Land Conservation Technologies in the Ethiopian Highlands: A Case Study in Andit Tid, North Shewa. *Agricultural Economics* 18: 233–47.

Shivley, G. 1999. Risk and Returns from Soil Conservation: Evidence from Low-Income Farms in the Philippines. *Agricultural Economics* 21(1): 53–67.

Soule, M. J., and K. D. Shepherd. 2000. An Ecological and Economic Analysis of Phosphorus Replenishment for Vihiga Division, Western Kenya. *Agricultural Systems* 64: 83–98.

Stewart, F. 2005. Groups and Capabilities. *Journal of Human Development* 6(2): 185–204.

Stiglitz, J., and A. Weiss. 1981. Credit Rationing in Markets with Imperfect Information. *American Economic Review* 71(3): 393–410.

Swinton, S. M., and R. Quiroz. 2003. Poverty and the Deterioration of Natural Social Capital in the Peruvian Altiplano. *Environment, Development, and Sustainability* 5: 477–90.

Tiffen, M., M. Mortimore, and F. Gichuki. 1994. *More People, Less Erosion: Environmental Recovery in Kenya.* Chichester, UK: John Wiley.

Wooldridge, J. M. 2002. *Econometric Analysis of Cross Section and Panel Data.* Cambridge, MA: MIT Press.

World Bank. 1994. *Kenya: Natural Resources Management Study.* Agriculture and Natural Resources Division. Washington, DC: The World Bank.

Yesuf, M. 2004. *A Dynamic Economic Model of Soil Conservation with Imperfect Market Institutions.* Doctoral Dissertation. University of Gothenburg, Sweden.

Young, A. 1997. *Agro-forestry for Soil Conservation.* Wallingford, UK: CAB International.

Zak, P. J., and S. Knack. 2001. Trust and Growth. *Economic Journal* 111: 295–321.

Appendix 3.1. Survey Questions Used to Extract Social Capital Information

Social Capital

Associations

In this section, I would like to ask you about the groups or organizations, networks or associations to which you or any members of your household belong. These could be formally organized groups or just informal groups of people who meet regularly to talk or do an activity.

C1. Do you or any member of the household belong to any organization or association?

1. YES.............. 2. NO.................[Read out the possible types from the list] (Farmers group, Traders and Business Association, Church, Soccer Club, Agricultural club, Credit/Finance group, Merry-Go-Round, Village committee, Burial committee, Political group, Cultural group, Water group, NGO, Civic group, and so on).

C2. Of all the groups to which members of your household belong to, which are the three (3) most important to you and/or your household?

a)   b)   c)

C3. How many times in an average month did anyone in the household participate in each of these groups' activities, e.g. by attending meetings and group work?

C4. How much money, time, or goods did your household contribute to the group last year?

4a. **Money** (amount Kshs)	4b. **Time** (hours)	4c. **Goods** (value Kshs)
Group 1: a)	b)	c)
Group 2: a)	b)	c)
Group 3: a)	b)	c)

C5. What are the *two* main benefits from joining the groups?
    For example, improved household access to livelihood and access to services, important in times of emergency, beneficial to the community, enjoyment/leisure, social status/self-esteem, others (please specify).

Group 1: a)	b)
Group 2: a)	b)
Group 3: a)	b)

C6. Does the group help your household with any of the following services? 1. YES 2. NO

	Group 1	Amount	Group 2	Amount	Group 3	Amount
Agricultural inputs (seed, pesticide, technical advice, etc.)						
Artificial insemination services						
Credit/savings services						
Soil conservation advice/information						
Information on crop prices/market opportunities						

Personal Friends and Contacts

T1. About how many *close* friends do you have these days? (These are people you feel at ease with, can talk to about private matters, or call for help.)

T2. If you suddenly needed a small amount of money—enough to pay for expenses for your household for one week—how many people beyond your immediate family could you turn to?

a) No one   b) One to two   c) Three to four   d) Five or more people
(Please tick one).

T3. Suppose you suffered a serious economic setback, such as crop loss. How many people could you turn to for help in this situation beyond your immediate family?

a) No one   b) One to two   c) Three to four   d) Five or more people
(Please tick one).

T4. In the past year, how many people with a personal problem have turned to you for assistance?

T5. If so, please state the value/amount _____ Kshs.

Neighborhood Relations

N1. Have you during the past year assisted anyone with significant amount of tools? (*jembe*, fork, hoes, wheelbarrows, spades etc.) 1. YES... 2. NO...

N2. Have you or household received such help? 1. YES .... 2. NO...........

N3/4. If a community project does not directly benefit you, but has benefits for many others in the neighborhood, would you contribute time or money to it?

TIME	MONEY
a) Will not contribute time [1]	a) Will not contribute money [1]
b) Will contribute time [2]	b) Will contribute money [2]

N5. In the past year, have you worked with others in the community/village to do something for the benefit of the community? 1. YES   2. NO

If Yes, please state the activity..........................

Sources of Market Information

I1-3. What are the *three* most important sources of market information (jobs, price of good or crops, etc.)?

a) Community centers, b) Relatives, friends, neighbors, c) Radio, d) Television e) Community leaders, f) NGOs g) Business associates, h) Groups/Associations, i) Government agents, j) Internet, k) National newspapers, l) Others

I4-6. What are the most important sources of information about what the government services (such as agricultural extension, tree planting week, family planning, etc.)?

a) Community centers, b) Relatives, friends, neighbors, c) Radio, d) Television e) Community leaders, f) NGOs g) Business associates, h) Groups/Associations, i) Government agents, j) Internet, k) National newspapers, l) Others

CHAPTER 4

Tenure Security and Incentives for Sustainable Land Management: A Case Study from Kenya

JANE KABUBO-MARIARA AND VINCENT LINDERHOF

C rop productivity growth, conservation, and sustainable utilization of the environment and natural resources are now recognized as critical national poverty reduction issues. Poverty Reduction Strategy Papers (PRSP), which in many countries are the key poverty reduction policy documents,[1] typically recognize that weak environmental management, unsustainable land use practices, and depletion of the natural resource base undermine crop productivity and reduce household welfare. The Kenyan PRSP, for example, recognizes the importance of sustainable land management, though the factors affecting adoption of land management technologies are less well known.

As in other sub-Saharan African countries, land degradation in Kenya is a major environmental concern and presents a formidable threat to sustainable agricultural production. Indeed, livelihoods are often bolstered by land management practices that contribute to soil erosion and other forms of land degradation, thereby jeopardizing sustainable development (Kabubo-Mariara et al. 2006). Agricultural potential, market access, and population pressure define important aspects of development, but less-favored areas are typically characterized by a combination of low agricultural potential, high population density, and poor market access. They often also have institutional settings that are not conducive to development.

Major institutional elements of development include property rights, which often have important implications for land management and sustainability. Property rights can affect land management through incentives for collective action and by affecting household incentives and abilities to invest (Pender et al. 2006; Besley 1995; Boserup 1965; Brasselle et al. 2002; Jacoby et al. 2002). As has been emphasized already and will continue to be discussed throughout this volume, household tenure rights may in particular have important implications for adoption of sustainable land management technologies (Kabubo-Mariara et al. 2006).

This chapter adds to the growing literature on sustainable land management by exploring the impact of tenure security and other development domains on adoption in Kenya. Using household and community survey data from Narok, Murang'a and Maragua districts, the chapter analyzes use of grass strips and terraces to prevent soil erosion. We address the following research question: What role do tenure security and other factors play in the adoption of land management technologies? The rest of the chapter is organized as follows. First, we present the methodology. The next section describes the sampling procedures, data, and summary statistics. We then present the results and offer conclusions.

Methodology

Before presenting our empirical specifications, we must lay out the conceptual framework hypotheses under which this assessment is conducted.

Conceptual Framework and Hypotheses

The conceptual framework utilizes theories of induced innovation in agriculture to explain management systems in terms of microeconomic incentives (Pender et al. 2006; Boserup 1965; Kabubo-Mariara 2005). According to Esther Boserup, as population grows, land and other natural resources become scarce relative to labor, while access to markets improves. As a result, agricultural intensification occurs, relative prices change, and food prices increase as demand increases. This process induces institutional innovations like private property that facilitate adoption of better technologies that stave off diminishing returns.[2] The same premise is held by evolutionary land rights theory (Platteau 1996, 2000).[3]

Following the literature we hypothesize that land tenure security is one of the major factors influencing adoption of land management practices. As others, including Wilfred Nyangena in Chapter 3, emphasize, biophysical factors determining agricultural potential, population density, access to markets, infrastructure and services, and endowments of physical, human, social, and natural capital should also be included in adoption analyses. Market access is hypothesized to have a direct effect on adoption of land management practices by increasing the profitability and availability of inputs and credit, though the sign of the impact is ambiguous. For example, market-driven intensification may lead to adoption and improved soil fertility, because of increased incentives to invest (Tiffen et al. 1994; Pender et al. 2006; Kabubo-Mariara et al. 2006). Market stimuli could also contribute to increased erosion, however, by reducing fallowing unless sufficiently offset by adoption of more intensive soil management and fertility practices.

Access to programs, services, and other institutions alleviate liquidity constraints and affect access to information about technologies, capacities to use technologies, and abilities to organize collective action (Pender et al. 2006). Technical assistance programs and organizations like agricultural and natural resource management extension groups may have significant influence on land management by enabling farmers to purchase inputs, labor, and capital.

The impact of land rights on management depends on the nature of tenure, and both private and common property may be important for land management decisions. In some cases the two regimes are closely linked. For instance, under private property, customary tenure often determines land use rights and management obligations of farmers, security of rights, whether they can be transferred or used as collateral, and how conflicts are resolved (Pender et al. 2006; Besley 1995; Boserup 1965; Brasselle et al. 2002; Jacoby et al. 2002).

Empirical Specification

The decision to adopt land management practices (LMP) is specified as:

$$LMP = f(PRR, Z, \varepsilon) \tag{4.1}$$

where PRR is a vector of tenure characteristics (mode of acquisition, use rights, bequest rights, etc.),[4] Z is a vector of conditioning variables (human, social and physical capital, plot characteristics, and village characteristics), and ε is a random error term. Estimation of equation (4.1) poses challenges due to potential endogeneity, because participation in programs and organizations may be endogenous to land management. Interest groups, extension agencies, and willingness to listen to those groups affect willingness to invest in land management.

To solve the endogenity problem, the best solution would be instrumental variables (IV) estimation based on Hausman tests of endogeneity and tests of exclusion restrictions (Wooldridge 2002). An alternative is to use a stepwise error correction approach. This involves the use of predicted values or residuals of the potentially endogenous variables as instrumental variables in the estimation of the truly endogenous variables (Pender and Gebremedhin 2006; Kabubo-Mariara et al. 2006; Nkonya et al. 2004). The resulting equation allows estimation of the direct and indirect effects of exogenous variables on dependent variables and also eliminates potential endogeneity bias. We use this approach in this chapter.

To capture the impact of interest group/village institution membership, we specify probit models for membership in income generation, loan and benevolent groups. We also run probit models for willingness to listen to extension and natural resource management agents. The residuals from each probit equation then enter equation (4.1) as error correction variables in order to yield unbiased estimates for the other coefficients. The coefficients of the independent variables capture both the direct and indirect effects (i.e., the impacts of exogenous variables on other decisions, such as participation

in interest groups), while the residual terms capture the effects of the endogenous variables (Kabubo-Mariara et al. 2006).

The limitation of this approach is that the probit models for groups and
extension services are identical equations. Seemingly unrelated regressions
would be the best estimation method, except our dependent variables are
discrete. We therefore estimate probit models and correct for correlation of
variances. We do this by applying factor analysis to the residuals of each probit
model to find common variance factors. These factors are then included in the
final model of willingness to invest in land management practices (Kabubo-
Mariara et al. 2006).

Sampling Procedures, Data and Summary Statistics

Before presenting the data, we will enumerate the sites themselves.

Study Site and Sampling

This sample is made up of 684 plots drawn from a self-weighting probability
sample of 457 households. The data were collected in November and December
2004 from Murang'a (151), Maragua (188), and Narok (118) districts. The
National Sample Survey and Evaluation Programme (NASSEP[5]) IV of the
Central Bureau of Statistics, Ministry of Planning, and National Development
is used as the sampling frame for the field survey.

The first stage in the sampling procedure is to select districts based on
differences in poverty, population density, and terrain. Administrative divi-
sions are selected within each of the three districts based on agro-ecological
diversity. Locations and sub-locations are selected using similar criteria, and
clusters from the NASSEP frame are chosen based on the total number of clusters within sub-locations and number of households in each cluster.

In the final stage, households are selected from each cluster. In addition
to the household survey a community questionnaire is administered to key
informants in each cluster (village). The community survey assembles information on market access and infrastructure and is meant to supplement
information collected from households.

Sample Statistics

Variables to explain adoption of land management technologies include
tenure security, soil quality and topography, market access, and institutions.
We collect a variety of information, yielding a very large vector for each set of
factors. To reduce the number of explanatory variables and avoid arbitrariness
in choice of variables, we use factor analysis[6] applied to tenure security, soil
quality, market access, and institutional factors.

TABLE 4-1 Summary Statistics on Sustainable Land Management

Type of investment	Number of plots	Share of plots
Grass strips	230	0.34
Ridging	12	0.02
Fallowing	31	0.05
Stone terraces	1	0.001
Soil terraces	59	0.09
Terracing with hedges	59	0.09
Terracing with grass strips	254	0.37
None	48	0.04
Period of investment		
Last year	229	0.33
Last five years	189	0.28
More than five years ago	203	0.30
Already on land upon acquisition	64	0.09
Total number	684	1.00

Land management practices

The distribution of land management practices per plot and the period of investment are presented in Table 4-1. The most frequently observed land management investments are terracing with grass strips (37%) and grass strips only (34%), meaning 54% of all plots have at least one technology. The data indicate that 72% of investments are permanent, 33% were made within the last year, and 30% more than five years earlier. The econometric analysis focuses on adoption of grass strips and terracing.

Tenure security

To capture all aspects of tenure security, data are collected on the mode of acquisition, perceived land rights, and whether plots are cultivated or rented/lent out. The mode of acquisition focuses on how and when plots were acquired and for how long plots have been in the household. Land rights focus on the perception of whether land is shared, can be taken away, etc. In addition, we probe rental arrangements and rights on rented and lent land. A summary is presented in Table 4-2.

The data show that households own (and typically use) 71%, rent-in 22%, and rent out 7% of plots. More than half are inherited, average ownership is over 18 years, and 5% of plots are owned over 50 years. Forty-six percent of plots are registered in the name of the nuclear household head or spouse, while 31% are registered by relatives, parents and siblings.

TABLE 4-2 Summary Statistics on Tenure Security

Variable	Mean	Std. dev.
Ownership (dummies)		
Own plot	0.71	0.46
Rented plot	0.22	0.42
Plot rented out	0.07	0.26
Acquisition		
Purchased plot	0.10	0.30
Gifted plot	0.05	0.22
Inherited plot	0.62	0.49
Ownership duration		
Ownership duration in years	18.1	14.7
Ownership duration 50 years or more (dummy)	0.05	0.21
Registration (dummies)		
Plot registered to head or spouse	0.46	0.50
Plot registered to another relative	0.31	0.46
Permission (dummies)		
Sell without permission	0.36	0.48
Sell or bequeath with permission	0.09	0.28
Bequeath without permission	0.21	0.41
Rent or lent with(out) permission	0.12	0.33
Permission of a relative	0.10	0.30
Arrangement (dummies)		
Rental arrangement	0.14	0.34
Indefinite arrangement	0.26	0.44

The rotated factor loadings lead to the selection of five orthogonal factors that explain almost 80% of the variance in tenure security (Table 4-3). The first factor is referred to as *farmland*, which ranges from full ownership to indefinite rental arrangements. These plots are registered to either the household head or spouse. The second factor reflects plots owned by the family (*family land*). The third (*land for sale*) includes land for which relatives have to give permission to sell or bequest. The fourth factor captures rented land (*land rented out*), and the final one focuses on plots that are either rented or lent with or without permission (*rental land*).

Soil quality and topography

We also utilize data on soil quality, including soil type, workability, texture, depth, and fertility. Table 4-4 shows the summary statistics. In the survey 42% of plots have red soils, 23% black, 25% a mixture of red and black, and the

TABLE 4-3 Factor Analysis for Tenure Security at Plot Level

Factor	Eigenvalues	Difference between eigenvalues	Proportions of variance explained		Description of the factor
			Marginal	Cumulative	
1	5.123	2.139	0.319	0.319	Farmland
2	2.983	1.215	0.186	0.505	Family land
3	1.768	0.208	0.110	0.616	Land for sale
4	1.561	0.208	0.097	0.713	Rented out land
5	1.353	0.392	0.084	0.797	Rental land

*Calculated using iterated principal factor method (see Kabubo-Mariara et al. 2006 for more details).

TABLE 4-4 Summary Statistics on Soil Characteristics (Plot Level)

Characteristics	Number of plots	Share of plots
Soil type		
Red	287	0.42
Mixed	174	0.25
Black	157	0.23
Rocky	51	0.07
White	15	0.02
Slope		
Flat	150	0.22
Weak undulation	78	0.11
Slightly sloped	256	0.37
Moderately sloped	101	0.15
Steeply sloped	99	0.14
Workability		
Easy	409	0.60
Moderate	186	0.27
Difficult	89	0.13
Soil texture of the plot		
Coarse	98	0.14
Intermediate	257	0.38
Fine	327	0.48
Perceived soil quality		
Very fertile	76	0.11
Fertile	266	0.39
Average	255	0.37
Not fertile	66	0.10
Very poor	16	0.02

TABLE 4-5 Factor Analysis Results for Soil Quality and Topography

Factor	Eigenvalues	Difference between eigenvalues	Proportions of variance explained		Description of the factor
			Marginal	Cumulative	
1	3.617	1.422	0.151	0.151	Texture
2	2.195	0.135	0.092	0.242	Fertility
3	2.060	0.237	0.086	0.328	Unknown fertility
4	1.823	0.328	0.076	0.404	Difficult workability
5	1.495	0.074	0.062	0.466	Flat vs. slight slope
6	1.421	0.070	0.059	0.525	Very fertile
7	1.351	0.078	0.056	0.582	Coarse soil
8	1.274	0.115	0.053	0.635	Moderate slope
9	1.158	0.044	0.048	0.683	Red vs. black soil
10	1.115	0.103	0.046	0.730	Undulated
11	1.011	0.033	0.042	0.772	Poor soil

*Calculated using iterated principal factor method (see Kabubo-Mariara et al. 2006 for more details).

remainder other soils. Two-thirds are slight-to-steeply sloped, while 22% are flat. The workability on 60% of plots is easy, texture on almost half is fine, and only 14% have coarse textures; average depth is 22.5 cm. Respondents said 12% of plots had poor or very poor fertility and 11% very high fertility.

Eleven factors related to soil quality and topography are selected from the factor analysis (Table 4-5). The first two reflect texture (fine to intermediate) and fertility (modest to average). The third and fourth factors indicate unknown fertility and difficult workability. The fifth factor represents plot slope (flat to slight slope) and the rest include very fertile, poor, coarse, and red soil, with undulated and moderate slopes.

Institutional factors

Institutional variables capture both presence and membership in village institutions. Factor analysis is based on the number of institutions in each community, 27 dummy variables for type of organization (village, men's, women's, and other groups) combined with purpose (livestock, agriculture, burial and illness, income generation, investments, natural resource management). The factor analysis yields 5 variables reflecting the presence of men's, income generating, village, safety net, natural resource management, and other groups.

Market access

Market access variables include distance, mode, travel time and expenses from the village to destinations like markets, roads, and other facilities. This information is collected using a community questionnaire. The factor analysis for market access is limited to distance to local market, all-weather roads, public transport (*matatu*), and nearest town. This analysis yields one factor reflecting travel expenses per kilometer as the best measure of market access.

Determinants of Land Management Adoption

Probit results for adoption of land management practices are presented in Table 4-6.[7] Tenure security, which is our primary interest, is captured by owned plots, family land that can be sold or bequeathed with or without permission, land registered in the family name, rented out land, and the right to rent out without permission. The estimated coefficients for the first three variables, which represent the strongest rights, are typically positive and also significant for land registered to household heads and spouses, suggesting the importance of tenure security in land management investments; having land registered promotes all investments analyzed. The negative and mostly significant coefficients of the last two variables indicate that land that is rented in or out or lent tends to not have investments.

The results suggest that few household characteristics influence adoption. While this result is not uncommon in the literature (e.g., Gebremedhin and Swinton 2003; Kabubo-Mariara et al. 2006; Kabubo-Mariara 2007), it could also be because our models capture both direct and indirect effects. The number of children aged between 6 and 16 years is positively correlated with adoption of all types, suggesting that older children may provide important labor inputs. The number of very young children (< 5 years old) is negatively correlated with adoption of terracing, but positively correlated with all practices.

Market access and population density at the village level are positively correlated with adoption of terraces, but insignificant for grass strips and all investments.[8] The positive coefficient on the Murang'a dummy suggests a higher probability of adoption if a household is located in Murang'a rather than Maragua district, and this is especially true for permanent investments. This likely reflects unobservable spatial differences (Kabubo-Mariara et al. 2006). There does not appear to be an impact of plot size on adoption, but distance to plot is negatively correlated with terraces, implying that terraces are more likely to be adopted on home rather than distant plots. Few agricultural potential variables are significant, though moderate slope favors adoption in all models. Assets also appear to play a limited role in adoption, though for all investments, livestock value is positively correlated with adoption. Existing conservation structures are negatively related to investments, indicating investments are made on plots without prior improvements. Results for village institutions are inconclusive, with membership in some institutions promoting land management technologies and others discouraging adoption.

TABLE 4-6 Determinants of Land Management Adoption

	Grass strip terraces	All terraces	All investments
Tenure security and related factors			
Land registered to household head or spouse	0.1714 [1.93]*	0.3425 [4.11]***	0.1856 [2.80]***
Family land registered in extended family	−0.0237 [0.31]	0.0689 [0.95]	0.0322 [0.54]
Right to sell family land with permission	−0.0757 [0.87]	0.2607 [3.93]***	−0.0634 [0.99]
Rented land	−0.1825 [1.38]	−0.3406 [2.92]***	−0.2664 [2.90]***
Lent out land	−0.2335 [2.04]**	−0.1421 [1.55]	−0.1378 [1.95]*
Household characteristics			
Child less than age 5 in a household	0.0232 [0.21]	−0.2075 [2.01]**	0.1378 [1.82]*
Children ages 6 to 16 in a household	0.0346 [0.55]	0.0918 [1.61]	0.0004 [0.01]
Number of adults in a household	−0.0942 [1.05]	0.0458 [0.63]	−0.0081 [0.14]
Household head years of schooling	−0.0148 [0.62]	0.0049 [0.23]	0.0231 [1.37]
Village Characteristics			
Number of institutions present	0.7096 [2.75]***	−0.0044 [0.02]	0.1633 [1.24]
Presence of men's groups	0.0788 [0.78]	−0.0013 [0.01]	−0.0286 [0.43]
Presence of income generating groups	0.3436 [1.94]*	−0.0747 [0.52]	0.1222 [1.34]
Presence of village committees/ groups	0.2109 [1.33]	0.0045 [0.03]	−0.0272 [0.28]
Presence of safety net and NRM groups	−0.6346 [2.78]***	−0.6875 [2.89]***	−0.1522 [1.24]
Population density	−0.0005 [0.61]	0.002 [2.41]**	−0.0002 [0.43]
Market access	0.1509 [0.48]	0.4488 [1.87]*	0.1081 [0.70]
Murang'a district dummy	0.1761 [0.72]	0.3794 [1.69]*	0.7729 [4.05]***
Soil quality and Topography			
Moderate texture	−0.0033 [0.04]	0.0081 [0.10]	−0.357 [4.80]***
Very fertile soils	−0.2287 [1.43]	0.0927 [1.08]	−0.0801 [1.13]
Fertile to average fertile	0.0725 [0.81]	−0.0962 [1.12]	−0.1182 [1.72]*

TABLE 4-6 *(Cont.)*

	Grass strip terraces	All terraces	All investments
Coarse soils	−0.0746	0.0804	−0.0809
	[0.63]	[0.93]	[1.13]
Red vs. black soils	−0.1376	−0.082	−0.0585
	[1.35]	[0.93]	[0.77]
Poor soils	−0.0643	−0.0132	0.052
	[0.73]	[0.17]	[0.82]
Soil depth	−0.0057	0.0196	0.0183
	[0.36]	[1.42]	[1.76]*
Steep slope	0.0625	−0.013	0.0032
	[0.79]	[0.17]	[0.05]
Moderate slope	0.3001	0.2934	0.1936
	[3.50]***	[3.60]***	[3.05]***
Undulating terrain	−0.0017	−0.2095	0.0171
	[0.02]	[2.52]**	[0.26]
Flat slope	−0.1249	0.1515	−0.0354
	[1.24]	[2.14]**	[0.56]
Assets			
Plot area (farm size)	−0.0003	−0.0019	0.0081
	[0.02]	[0.17]	[1.21]
Distance to plot	−0.0106	−0.0254	−0.0038
	[1.44]	[2.60]***	[0.82]
Lagged value of livestock (log)	0.0957	0.035	0.1713
	[0.99]	[0.41]	[2.51]**
Lagged value of farm equipment	0.0526	0.1024	−0.0146
(log)	[0.77]	[1.62]	[0.29]
Previous soil conservation	−0.6589	−1.124	−0.4512
structures	[4.90]***	[7.68]***	[5.27]***
Error correction terms (residuals)			
Listened to extension services	0.1123	0.0836	−0.0561
	[0.83]	[0.70]	[0.61]
Membership in village	−0.0996	−0.1691	−0.0484
institutions	[0.92]	[1.70]*	[0.61]
Willingness to invest in SWC	0.3411	0.5728	0.6487
	[2.07]**	[3.79]***	[5.33]***
Constant	−1.2782	−2.6468	−1.6975
	[2.22]**	[4.74]***	[3.98]***
Observations	684	684	684
LR chi2(36)	94.04***	204.87***	101.56***
Pseudo R2	0.2396	0.3534	0.1533
Log likelihood	−149.18	−187.393	−280.4

Absolute value of z statistics in brackets
* significant at 10%; ** significant at 5%; *** significant at 1%

Conclusions

This chapter investigates the impact of tenure and other factors on adoption of land management practices. Factor analysis is employed to arrive at the final variables for tenure security, village level institutions, agricultural potential and market access. Results largely affirm the importance of tenure security. Indeed, the analysis suggests that the more secure farmers are about their land rights, the more likely they are to adopt land management practices. Favorable agro-ecological potential, market access, and population density may also have positive impacts, but in contrast to the findings of Nyangena in Chapter 3, our analysis indicates that village institutions have little effect on adoption. These results suggest that to promote better land management the primary focus should be on tenure security.

Notes

1   For a complete list of published Poverty Reduction Strategy Papers, please see www.imf.org/external/NP/prsp/prsp.asp.
2   Although Boserup's hypothesis has been found to be consistent with certain empirical evidence from developing countries (Tiffen et al. 1994, among other studies), we note that it has been contradicted by some studies showing that population growth is associated with reduced productivity and environmental quality (Place and Otsuka 2002).
3   The evolutionary land rights theory contends that as land scarcity increases, people demand more land tenure security. As a result, private property rights in land tend to emerge; once established, they evolve toward greater measures of individualization and formalization.
4   Tenure variables are treated as exogenous. Though some literature suggests that investments in land management may affect tenure security, previous studies in Kenya indicate this is not the case (Kabubo-Mariara 2005, 2007).
5   The NASSEP frame has a two-stage stratified cluster design for the whole country. Enumeration areas are selected using the national census records, with the probability proportional to size of clusters. The number of clusters is obtained by dividing each primary sampling unit into 100 households. The clusters are then selected randomly.
6   For the selection of key variables from the factor analysis we use two criteria: (i) eigenvalues have to be larger than one, and (ii) the share of variance explained has to be at least 75%. The factors derived from factor analysis are orthogonal, ensuring that the final estimates are unbiased, consistent, and efficient.
7   The choices of terracing are likely to be related, and it would be appropriate to use a multinomial or multivariate model. However, the data do not allow adequate multivariate or multinomial analysis.
8   The dummy variable for Narok is dropped, because it is found to be highly correlated with population density. We test for multicollinearity between population density, market access, and the Murang'a district dummy, but the results suggest very low and insignificant correlation.

References

Besley, T. 1995. Property Rights and Investment Incentives: Theory and Evidence from Ghana. *Journal of Political Economics* 103(5): 903–37.

Boserup, E. 1965. *The Conditions of Agricultural Growth: The Economics of Agrarian Change under Population Pressure*. New York: Aldine Press.

Brasselle, A. S., F. Gaspart, and J. P. Platteau. 2002. Land Tenure Security and Investment Incentives: Puzzling Evidence from Burkina Faso. *Journal of Development Economics* 67: 373–418.

Gebremedhin, B., and S. M. Swinton. 2003. Investment in Soil Conservation in Northern Ethiopia: The Role of Land Tenure Security and Public Programs. *Agricultural Economics* 29: 69–84.

Jacoby H. G., G. Li, and S. Rozelle. 2002. Hazards of Expropriation: Tenure Insecurity and Investment in Rural China. *American Economic Review* 92(5): 1420–77.

Kabubo-Mariara, J. 2005. Herders Response to Acute Land Pressure under Changing Property Rights: Some Insights from Kenya. *Environment and Development Economics* 10(1): 67–85.

———. 2007. Land Conservation and Tenure Security in Kenya: Boserup's Hypothesis Revisited. *Ecological Economics* 64: 25–35.

Kabubo-Mariara J., V. Linderhof, G. Kruseman, R. Atieno, and G. Mwabu. 2006. Household Welfare, Investment in Soil and Water Conservation, and Tenure Security: Evidence from Kenya. Poverty Reduction and Environmental Management (PREM) Working Paper 06–06.

Nkonya, P., J. Pender, P. Jagger, D. Sserunkuuma, C. K. Kaizzi, and H. Ssali. 2004. Strategies for Sustainable Land Management and Poverty Reduction in Uganda. Research Report No. 133. Washington, DC: International Food Policy Research Institute.

Pender J., S. Ehui, and F. Place. 2006. *Strategies for Sustainable Land Management in the East African Highlands*. Washington, DC: International Food Policy Research Institute.

Pender J., and B. Gebremedhin. 2006. Land Management, Crop Production, and Household Income in the Highlands of Tigray, Northern Ethiopia: An Econometric Analysis. In *Strategies for Sustainable Land Management in the East African Highlands,* edited by J. Pender, S. Ehui, and F. Place. Washington, DC: International Food Policy Research Institute.

Place F., and K. Otsuka. 2002. Land Tenure Systems and their Impacts on Productivity in Uganda. *Journal of Development Studies* 38(6): 105–28.

Platteau, J. P. 1996. The Evolutionary Theory of Land Rights as Applied to Sub-Saharan Africa: A Critical Assessment. *Development and Change* 27(1): 29–86.

———. 2000, *Institutions, Social Norms, and Economic Development*. Amsterdam: Harwood Academic Publishers.

Tiffen M., M. Mortimore, and F. Gichuki. 1994. *More People, Less Erosion: Environmental Recovery in Kenya*. Chichester, UK: John Wiley and Sons.

Wooldridge, J. M. 2002. *Econometric Analysis of Cross Section and Panel Data*. Cambridge, MA: MIT Press.

The Effect of Risk on Investments

PART II

The Effect of Risk on Investments

CHAPTER 5

Risk Preferences and Technology Adoption: Case Studies from the Ethiopian Highlands

MAHMUD YESUF AND HAILEMARIAM TEKLEWOLD

F arm households in most developing countries face a multitude of risks, including those such as poor weather, price fluctuations, and pests. In many of these countries, credit and insurance markets are incomplete or missing, implying that households cannot pass risks to third parties. In the presence of such market failures, endowments determine many of the risk responses that affect farm investment decisions (Pender et al. 2001; Hagos 2003).

As discussed with regard to crop biodiversity by Di Falco and Chavas in Chapter 7, the impact of risk on the decision to adopt and use farm technologies critically depends on the risk properties of those technologies (Just and Pope 1978, 1979) and the degree of consumption risk facing households (Shively 1997). When technologies are risk-increasing, risk-averse households are less likely to adopt, but the opposite is true when technologies reduce risk; households may also adopt if they have no other choice.

Despite the likely importance of risk responses, few studies have measured farm household risk aversion and analyzed potential effects on farm investments. Using data from eastern Amhara, Yesuf (2004) finds a negative relationship between risk aversion and fertilizer adoption, suggesting that fertilizer is risk-increasing. In Chapter 6, as well as in past work (e.g., Hagos and Holden 2001), Fitsum Hagos and Stein Holden find a positive correlation between risk aversion and fertilizer adoption in Tigray, indicating the opposite relationship. Such results suggest the need for more investigation. This chapter therefore examines the effect of risk aversion on the use and application of chemical fertilizers in highland Ethiopia.

With the exception of Yesuf (2004) and Yesuf and Bluffstone (2009), all studies use hypothetical experiments to measure risk attitudes, which can yield hypothetical bias. This chapter uses real payoff experimental data to measure farmers' risk aversion. We then apply those findings to the problem of understanding the role of risk preferences in chemical fertilizer adoption in the Ethiopian highlands. The experiments are conducted in three different areas—East Gojam, South Wollo, and Northern Shewa—which have similar agro-ecological and resource endowments. All three zones have rugged terrain and erratic but intensive rain that can cause severe soil erosion. East Gojam and Northern Shewa are relatively high-potential areas, but South Wollo has recurrent drought and even famines.

In order to enhance the efficacy of our estimates, a two-stage estimation procedure is employed. First, a random effects linear model is used to analyze the determinants of risk aversion. The fitted values are then saved and used in a second stage model. In the second stage a Heckman selection model with random effects probit and linear specifications is employed to analyze the impact of risk preferences on adoption and intensity of chemical fertilizer use.

In the following section we describe our theoretical and empirical approach and risk aversion experiment. In the subsequent section we present the data and discuss the study sites. Next, we present the experimental findings and determinants of household risk aversion. We then describe our econometric models of fertilizer adoption and results before concluding the chapter.

Measuring Risk Aversion

Risk aversion is defined with reference to the von Neumann-Morgenstern expected utility function, with the second derivative of this utility function containing important information about risk aversion. In empirical studies experimental and production function approaches have been used to measure risk aversion. The experimental approach utilizes hypothetical questionnaires focusing on risky alternatives or games with or without real payments. Studies using this method in developing countries include Binswanger (1980), Wik and Holden (1998), and Yesuf (2004). The production function approach is based on actual production data (Antle 1987) and has been criticized for confounding risk behavior with factors like resource constraints. This may be particularly true in poor countries where market imperfections are prominent and production and consumption decisions unseparable (Wik and Holden 1998).

In this chapter an experimental method similar to that of Binswanger (1980) is employed to elicit farmers' risk preferences, but our experiments use real payoffs and mimic farming situations. The basic structure of the experiment is described in Table 5-1 and is replicated four times using the factors 0.05, 0.25, 0.5, and 1.5 to evaluate if risk aversion changes when stakes change. Though the amounts may seem low, it must be recalled that incomes in the study area are very low, so the amounts listed indeed provide incentives for respondents to reveal their true preferences.

In the experiment farmers are confronted with risky experimental agricultural systems in which bad and good harvests each occur with 50% probabilities based on coin tosses. On average, each household won ETB 30 (30 Ethiopian *birr*), which is about 10% of an unskilled laborer's monthly income.

TABLE 5-1 The Payoffs and Classification of Risk Aversion

Choice	Payoffs (Birr)		Expected Gain (E)	Standard Deviation (SE)	Trade-offs (Z)*	Approximate Risk Aversion Coefficients (q)	Risk Aversion Class
	Bad harvest	Good harvest					
1	10.00	10.00	10.00	0.00	0.78 – 1.00	∞ to 7.47	Extreme
2	9.00	18.00	13.50	4.50	0.71– 0.78	7.47 to 1.74	Severe
3	8.00	24.00	16.00	8.00	0.50 – 0.71	1.74 to 0.81	Intermediate
4	6.00	30.00	18.00	12.00	0.33 – 0.50	0.81 to 0.32	Moderate
5	2.00	38.00	20.00	18.00	0.00 – 0.33	0.32 to 0.00	Slight
6	0.00	40.00	20.00	20.00	−∞ – 0.00	0.00 to −∞	Neutral

*Z is the trade-off between expected gains and standard deviations of two games ($Z = dE/dSE$)

To compute our measure of risk aversion, we employ a constant partial risk aversion (CPRA) utility function of the form $U = (1 - \gamma)c^{(1-\gamma)}$, where γ is a coefficient of risk aversion, and c is the certainty equivalent of a prospect. If a respondent is indifferent between two consecutive prospects (say 1 and 2) given that both prospects have equal probabilities of good or bad outcomes, then we have $E(U_1) = E(U_2)$, and $(1 - \gamma)c_1^{(1-\gamma)} = (1 - \gamma)c_2^{(1-\gamma)}$. The upper and lower limits of the CPRA coefficients for each prospect of our experiment are given in Table 5-1.

Data Sources and Experimental Sites

This study is based on data collected from two sets of experiments in Amhara Regional State. The first is part of a survey of 1,522 households done in 2002 in the northwestern Ethiopian highlands through research collaboration between Gothenburg University and Addis Ababa University and financed by Sida, the Swedish International Development Cooperation Agency. The experiment is administered to a random sample of 262 households in 5 districts of two different zones in Amhara Regional State (East Gojam and South Wollo). The second set of data is from a 2003 survey of 143 farm households in two districts of Northern Shewa Zone of the Amhara Regional State.

Virtually all households in the sample are subsistence farmers who rely on their farm production for all consumption needs. In such an environment, consumption and production decisions are made jointly, which implies that endowments, such as wealth, family size, and other household characteristics affect production outcomes (Jacoby 1993). Multi-stage random sampling is used for selecting peasant associations (villages) and households from each district. Data on a number of household and farm characteristics are also collected at plot and household levels. Descriptive statistics are presented in Table 5-2.

TABLE 5-2 Descriptive Statistics

| Variables | Northern Highlands (N = 262) | | | | | | | | Central Highlands | | | |
| | East Gojam | | | | South Wollo | | | | North Shewa | | | |
	Mean	Std	Min	Max	Mean	Std	Min	Max	Mean	Std	Min	Max
Gender of the head (1 = male)	0.96	0.19	0	1	1	0	1	1	0.90	0.31	0	1
Age of the head (years)	46.40	14.51	17	74	48.69	15.32	20	90	46.50	12.73	19	84
Education level of the head/literacy	0.258	0.44	0	1	0.29	0.46	0	1	0.56	0.50	0	1
Family size	5.58	2.31	1	13	6.22	2.71	1	15	5.69	2.18	1	11
Household dependency ratio (ratio of number of household members age < 15 to age >15)	1.05	0.72	0	3	1.11	0.89	0	5	1.14	0.77	0	4
Household farm size (hectares)	1.74	0.79	0.13	4	0.90	0.68	0.04	3.38	1	0.64	0.12	4
Plot size (hectares)	0.28	0.19	0.013	1.25	0.27	0.24	0.001	1.5	0.25	0.15	0.03	1
Value of domestic animals (proxy for stock of wealth) in *birr*, TLU* for N. Shewa	2733.51	1966.03	11	8870	1508.42	1311.99	4	6180	4.01	2.12	0	10.30
Annual liquid cash availability to a household (cash revenue less cash expenditure), *birr*/year	494.12	657.47	−2040.4	2460	398.59	1397.99	−1340	9565	71.50	657.76	−1933	2989
Use chemical fertilizers (1 = yes)	0.46	0.50	0	1	0.21	0.41	0	1	0.37	0.48	0	1
Intensity of use of chemical fertilizers (kg/ha) or kg/plot for N. Shewa	30.20	25.91	0	150	21.57	17.58	2.5	100	17.33	29.36	0	150
Use any SWC measure (1 = yes)	0.19	0.39	0	1	0.46	0.50	0	1	0.48	0.50	0	1
Use manure (1 = yes)	0.13	0.34	0	1	0.28	0.45	0	1	0.22	0.41	0	1
Soil quality (1 = poor)	0.28	0.45	0	1	0.12	0.33	0	1	0.44	0.50	0	1
Slope (1 = steep)	0.29	0.46	0	1	0.31	0.46	0	1	0.39	0.49	0	1
Distance of plot from home (in min.)	15.72	17.43	0	150	18.13	18.69	0	150	18	17.11	1	90
Constant partial risk aversion coefficient	2.69	2.67	0	7.47	3.70	2.98	0	7.47	3.22	2.94	0	7.50

*TLU = tropical livestock units

Experimental Results and Determinants of Household Risk Aversion

The distribution of risk averting behaviors corresponding to the experiment in the three study areas is presented in Table 5-3. The results reveal that, while there are significant differences across zones, in all study areas more than 30% of farm households fall into the severe to extreme risk aversion classes. Households in Northern Shewa (about 53.2% in severe and extreme categories) and South Wollo (50.1%) are more risk averse than households in East Gojam (30.1%), and less than 15% in those zones exhibit slight risk aversion or risk neutrality. This level compares with slight risk aversion or risk neutrality of about 33% in East Gojam. Intermediate and moderate risk averse household categories are similar across the three zones at 32% to 37%. Comparing the distribution of risk preferences to other studies in developing countries, Binswanger (1980) and Wik and Holden (1998) find that 83% of respondents in India and 52% in Zambia are in the intermediate to moderate risk category; risk aversion is therefore higher in highland Ethiopia than in the areas studied in India and Zambia.

These findings raise the question of why some households are more risk averse than others. Indeed, if all households have equal access to credit and insurance, we would not expect to observe such variation, which suggests access differences. This then leads to the conclusion that without well-functioning capital and insurance markets, household endowments are likely to be important factors governing household responses to risk (Yesuf and Bluffstone 2009; Pender et al. 2001; Hagos 2003).

Using a constant partial risk aversion utility function, risk aversion coefficients are imputed for each risk category. Because respondents are asked five risk preference questions, up to five risk aversion coefficients are calculated for each respondent. To analyze the determinants of risk preferences, we run a random effects linear model.[1]

Based on earlier empirical studies and the economic theory of market imperfections (e.g., Jacoby 1993), explanatory variables were selected. To

TABLE 5-3 Frequencies Distributions of Risk Categories

Risk Aversion	Northern Highlands (N = 262)				Central Highlands (N = 143)	
Class	East Gojam		South Wollo		Northern Shewa	
	%	Cumulative,%	%	Cumulative,%	%	Cumulative,%
Extreme	12.8	12.8	29.1	29.1	31.9	31.9
Severe	17.3	30.1	21.0	50.1	21.3	53.2
Intermediate	18.7	48.8	22.3	72.4	14.9	68.1
Moderate	18.2	67.0	12.9	85.3	17.0	85.1
Slight	16.3	83.3	6.6	91.9	4.3	89.4
Neutral	16.7	100.0	8.1	100.0	10.6	100.0

capture the effect of wealth on risk aversion, variables such as value of livestock and farm size—which are common measures of wealth in the study sites—and current cash availability and participation in off-farm work (proxies for liquidity constraints) are included. It has been suggested that market imperfections severely constrain substitution between different forms of wealth (Reardon and Vosti 1995; Holden et al. 1998). Under such conditions each wealth category will have an independent effect on risk aversion, so we include them individually rather than aggregating. Household characteristics like gender, age, education of respondents, family size, and number of dependents are also included. The results of our random effects model are given in Table 5-4.

Most of the variables in the regressions are significant across the three zones. As expected, livestock value is highly significant, with less risk aversion

TABLE 5-4 Estimates of the Random Effects Linear Model: Determinants of Risk Aversion

Variable	Northern Highlands		Central Highlands
	East Gojam	*South Wollo*	*Northern Shewa*
Sex of the head (1 = male)	1.899***	6.666***	0.598
	(0.365)	(0.662)	(0.417)
Age of the head, years	0.025***	0.012	0.002
	(0.006)	(0.013)	(0.010)
Education (1 = literate)	−0.149	0.067	−0.252
	(0.160)	(0.286)	(0.253)
Family size	−0.243***	−0.115*	−0.031
	(0.053)	(0.068)	(0.062)
Dependency ratio	−0.624***	−0.384*	0.147
	(0.128)	(0.201)	(0.170)
Farm size	−0.364***	−1.609***	−0.115
	(0.099)	(0.255)	(0.078)
Value of capital stock (*birr*)	−0.0005***	−0.0016***	−1.132***
	(0.0003)	(0.0001)	(0.061)
Cash liquidity	−0.0005***	−0.0001	−0.001**
	(0.0001)	(0.0001)	(0.0001)
Off-farm work (1 = yes)	0.0003**	0.0002	0.0004*
	(0.0002)	(0.0002)	(0.0003)
Constant	3.475***	6.666***	7.084***
	(0.484)	(0.662)	(0.668)
Number of observations	710	289	597
Rho	0.73	0.69	0.97
Wald χ^2	987.22	444.09	156.81

Dependent variable is constant partial risk aversion coefficient
*, **, *** refer to significance at 10%, 5% and 1% level respectively; figures in parentheses are standard errors.

observed when higher endowments are present. Land size is important in East Gojam and South Wollo, but not in the Central Highlands, while liquidity measured by off-farm work and current cash income affects risk aversion in East Gojam and Northern Shewa, but not South Wollo. While current cash income reduces risk aversion, access to off-farm activities seems to increase risk averting behavior. This could possibly be because households participating in off-farm work are often also the poorest. Effects of household characteristics are mixed. Education is not significant across the three zones. Male-headed households are more risk averse in East Gojam and South Wollo, but gender is insignificant in Northern Shewa. Age increases risk aversion only in East Gojam, as does family size and number of dependents in East Gojam and South Wollo.

Risk Preferences and Fertilizer Adoption

We now examine whether there is any relationship between the imputed risk coefficients and use of chemical fertilizer. In the literature, farm investment decisions are modeled either in profit or utility maximizing frameworks, subject to constraints. In the presence of perfect markets it is optimal for farmers to use technologies until marginal returns equal marginal costs, or where the marginal value/cost ratio (VCR) equals one. An implication of this result is that in the presence of complete markets household resource endowments have no role in farm investments.

As has been emphasized in this chapter and in other chapters in this volume, in most developing countries markets are far from perfect for many reasons, including imperfect information, transportation systems, and transaction costs. In some cases markets are completely absent, which can increase risk. In the presence of risk and uncertainty due to missing capital and insurance markets a VCR > 1 is therefore typically required for farm households to adopt technologies like fertilizer; profitability is therefore a necessary, but not sufficient condition for adoption. In addition to the nature of market failures, the risk implications of technologies like fertilizer also become important for adoption. Risk attitudes are affected by household endowments and other characteristics, which will therefore also influence adoption.

In this study we use a two-stage Heckman selection model to analyze the decision to adopt and use chemical fertilizers. Since our data are at plot level, we exploit the panel features by running a first-stage random effects probit model and a second-stage linear random effects (with inverse mills ratio included as an explanatory variable) model. Since some households have only one plot, we are not able to exploit the added advantages of fixed effects models. Important conditioning variables for both stages are plot characteristics, which are selected based on the adoption literature and our previous work. In order to address the potential endogeneity of risk aversion, we use predicted risk coefficients in both equations from the specification in

Table 5-4. Since predicted instead of observed values are used, we bootstrap the standard errors. Furthermore, in order to reduce the potential endogenity of manure and soil conservation variables, we include as many covariates as possible that are correlated with both variables. The parameter estimates of the random effects probit for adoption and random effects linear model for intensity are presented in Table 5-5.

TABLE 5-5 Heckman Selection Model: Determinants of Fertilizer Adoption

Variables	Northern Highlands		Central Highlands
	East Gojam	South Wollo	Northern Shewa
First-stage estimates: random effects probit			
Use any SWC measure (1 = yes)	−0.267*	0.375*	−0.049
	(0.156)	(0.222)	(0.143)
Use manure (1 = yes)	−0.624***	−0.154	−1.669***
	(0.168)	(0.233)	(0.259)
Plot size, ha	0.503	1.719***	6.137**
	(0.318)	(0.452)	(2.557)
Soil quality (1 = poor soil)	0.315**	0.182	−0.185
	(0.135)	(0.334)	(0.139)
Slope (1 = steep)	−0.188	−0.469**	0.175
	(0.131)	(0.241)	(0.144)
Distance of plot from home (in min.)	0.001	−0.008	0.003
	(0.003)	(0.006)	(0.004)
Education (1 = literate)	−0.047	−0.462*	−0.192
	(0.157)	(0.250)	(0.156)
Farm size	−0.198*	−0.290	0.539
	(0.109)	(0.223)	(0.489)
Cash liquidity	−0.0001	0.0001	−0.0001
	(0.001)	(0.0001)	(0.0001)
Off-farm work (1 = yes)	0.0002	−0.0002	−0.00004
	(0.0001)	(0.0001)	(0.0002)
Constant partial risk aversion coefficient	−0.191***	−0.084	−0.390***
	(0.043)	(0.096)	(0.039)
Constant	0.708*	−0.468	1.345***
	(0.292)	(0.597)	(0.275)
Number of observations	703	277	597
Rho	0.24	0.17	0.11
Wald χ^2	50.52	25.20	136.02
Second-stage estimates: random effects linear model: determinant of intensity of fertilizer use			
Use any SWC measure (1 = yes)	−8.162	−3.081	2.281
	(5.978)	(4.604)	(1.944)
Use manure (1 = yes)	−26.109***	8.384	−14.487***
	(8.753)	(5.265)	(2.610)

TABLE 5-5 *(Cont.)*

Variables	Northern Highlands		Central Highlands
	East Gojam	South Wollo	Northern Shewa
Plot size, ha	49.857***	23.997*	71.341*
	(5.425)	(13.990)	(38.367)
Soil quality (1 = poor soil)	8.905*	9.351	–0.929
	(5.542)	(7.668)	(1.893)
Education (1 = literate)	1.249	5.220	2.964
	(4.415)	(7.430)	(2.291)
Farm size	–8.450**	–5.939	–70.260***
	(4.393)	(5.788)	(6.727)
Cash liquidity	–0.002	–0.008*	0.003**
	(0.004)	(0.004)	(0.001)
Off–farm work (1 = yes)	0.008	0.008	–0.007***
	(0.004)	(0.005)	(0.002)
Constant partial risk aversion coefficient	–8.392***	–4.732***	–0.843
	(2.845)	(1.583)	(0.948)
Inverse mills ratio	92.569***	36.129	–39.630***
	(34.362)	(30.592)	(8.003)
Constant	64.934***	33.163***	4.005
	(15.881)	(12.233)	(5.646)
Rho	0.27	0.76	0.10
Wald χ^2	22.07	43.09	306.42

*, **, *** refers to significance at 10%, 5% and 1% level respectively; figures in parenthesess are bootstrapped standard errors. All household characteristics were used to predict risk coefficients and hence not included in our Heckman selection model.

Our variable of interest is risk aversion, and the negative effects on fertilizer use seem to suggest that chemical fertilizers are risk-increasing. This finding is consistent with similar studies in other countries that found chemical fertilizers are risk-increasing (see Just and Pope 1979; Rosegrant and Roumasset 1985; Roumasset et al. 1989; Smith et al. 1989; Pandey 2004). Agronomic research also suggests that fertilizer use increases the variance of crop yields, particularly when rainfall is erratic as in our study sites (e.g., Seligman et al. 1992).

We find that risk is highly significant in explaining adoption and intensity of chemical fertilizer use in two of the three zones and in explaining both decisions in East Gojam. In Northern Shewa risk is a significant determinant of adoption, but not intensity; the reverse is true in South Wollo. We have also included a number of wealth variables in the adoption and intensity equations, but their effects are weak and seem to affect fertilizer use through risk aversion. This finding is consistent with those of Yesuf and Bluffstone (2009).

The effects of plot characteristics on adoption vary across the three zones. Plot size is a significant determinant of adoption only in South Wollo and

Northern Shewa, but it is significant across the three zones in determining how much fertilizer is applied. Soil quality is significant in explaining both adoption and intensity only in East Gojam, and slope is significant in explaining adoption only in South Wollo. Chemical fertilizer and soil and water conservation adoption seem to be substitutes in East Gojam, but complements in South Wollo. Likewise, manure and chemical fertilizer are substitutes in East Gojam and Northern Shewa, where there is a relatively large concentration of livestock. At least in our study sites distance from homestead is not significant in explaining adoption decisions.

Conclusions

Capital accumulation is critical to rural growth, but growth can only happen when farmers are willing to take reasonable risks. Adopting appropriate mean-increasing technologies is one strategy, but such adoption often does not take place in rural areas of developing countries. Low fertilizer adoption and use are just two examples.

This study uses data from two surveys conducted in the northern and central highlands of Ethiopia. The primary purpose of these surveys is to use experimental methods to evaluate farmer risk aversion. The relationship between endowments and risk aversion is examined, followed by an analysis of risk aversion and fertilizer use. We find that in the Ethiopian highlands risk aversion is very high by any standards. Livestock, which is the most common store of wealth, is found to be a significant determinant of risk behavior, which suggests that wealth and capital accumulation are perhaps the most direct way to reduce risk averting behavior in the presence of missing capital and insurance markets.

Our two-stage Heckman selection model results seem to suggest that, although plot characteristics are important depending on specific sites, risk preferences are by far the most important parameters affecting both the adoption and intensity of fertilizer use in areas where rainfall is highly erratic and farmers have repeated experience with drought and crop failures. Wealth appears to affect fertilizer use mainly through reducing risk aversion.

These findings suggest the need for policymakers to directly address the risk-increasing properties of fertilizers. Adopting and using risky mean-increasing investments like fertilizer is important for growth and increasing rural incomes. To avoid perpetuating poverty traps in environments where insurance and capital markets are missing, provision of appropriate insurance seems particularly important.

Notes

1   The random effects model has an advantage over other linear models as estimates are based on more information; the unobserved heterogeneities across households are controlled by the panel nature of the dependent variable.

References

Antle, J. M. 1987. Econometric Estimation of Producers' Risk Attitudes. *American Journal of Agricultural Economics* 69: 509–22.

Binswanger, H. 1980. Attitudes toward Risk: Experimental Measurement in Rural India, *American Journal of Agricultural Economics* 62(3): 395–407.

Hagos, F. 2003. Poverty, Institutions, Peasant Behavior, and Conservation Investment in Northern Ethiopia. Ph.D. Dissertation No. 2003:2. Department of Economics and Social Sciences, Agricultural University of Norway, Ås, Norway.

Hagos, F., and S. Holden. 2001. Incentives for Conservation in Tigray, Ethiopia: Findings from a Household Survey. Department of Economics and Social Sciences, Agricultural University of Norway. Unpublished.

Holden, S. T., B. Shiferaw, and M. Wik. 1998. Poverty, Market Imperfections, and Time Preferences: Of Relevance for Environmental Policy? *Environment and Development Economics* 3: 105–30.

Jacoby, H. 1993. Shadow Wages and Peasant Family Labor Supply: An Econometric Application to the Peruvian Sierra. *Review of Economic Studies*, 60: 901–23.

Just, R. E., and R. Pope. 1978. Stochastic Specification of Production Functions and Economic Implications. *Journal of Econometrics* 7: 67–86.

———. 1979. Production Function Estimation and Related Risk Consideration. *American Journal of Agricultural Economics* 61(2): 276–84.

Pandey, S. 2004. Risk and Fertilizer Use in the Rain-fed Rice Ecosystems of Tarlac, Philippines. *Journal of Agricultural and Applied Economics* 36(1): 241–50.

Pender, J., B. Gebremedhin, S. Benin, and S. Ehui. 2001. Strategies for Sustainable Development in the Ethiopian Highlands. *American Journal of Agricultural Economics* 83(5): 1231–40.

Reardon, T., and S. Vosti. 1995. Links between Rural Poverty and the Environment in Developing Countries: Asset Categories and Investment Poverty. *World Development*, 23(9): 1495–1506.

Rosegrant, M. W., and J. A. Roumasset. 1985. The Effect of Fertilizer on Risk: A Heteroskedastic Production Function with Measurable Stochastic Inputs. *Australian Journal of Agricultural Economics* 29: 107–21.

Roumasset, J. A., M. W. Rosegrant, U. N. Chakravorty, and J. R. Anderson. 1989. Fertilizer and Crop Yield Variability: A Review. In *Variability in Grain Yields: Implications for Agricultural Research and Policy in Developing Countries*, edited by J. R. Anderson and P. B. R. Hazell. Baltimore: Johns Hopkins University Press.

Seligman, N. G., H. van Keulen, and C. J. T. Spitters. 1992. Weather, Soil Conditions, and the Inter-Annual Variability of Herbage Production and Nutrient Uptake on Annual Mediterranean Grasslands. *Agricultural and Forest Meteorology* 57: 265–79.

Shively, G. 1997. Consumption Risk, Farm Characteristics, and Soil Conservation Adoption among Low-income Farmers in the Philippines. *Agricultural Economics* 17: 165–77.

Smith, J., G. M. Umali, A. M. Mandac, and M. W. Rosegrant. 1989. Risk and Nitrogen Use on Rain-fed Rice: Bicol, Philippines. *Fertilizer Research* 21: 113–29.

Wik, M., and S. Holden. 1998. Experimental Studies of Peasants' Attitudes toward Risk in Northern Zambia. Discussion Paper D-14. Department of Economics and Social Sciences, Agricultural University of Norway, Ås, Norway.

Yesuf, M. 2004. Risk, Time, and Land Management under Market Imperfections: Case Studies from Ethiopian Highlands. Ph.D. Thesis, Gothenburg University, Sweden.

Yesuf M. and R. A. Bluffstone. 2009. Poverty, Risk Aversion, and Path Dependence in Low-income Countries: Experimental Evidence from Ethiopia. *American Journal of Agricultural Economics* 91(4): 1022–1037.

CHAPTER 6

Fertilizer Use by Smallholder Households in Northern Ethiopia: Does Risk Aversion Matter?

FITSUM HAGOS AND STEIN HOLDEN

A s a number of previous chapters have noted, land degradation due to soil erosion and nutrient depletion poses serious challenges for farmers in the developing world. Ethiopia has one of the highest nutrient depletion rates in sub-Saharan Africa (Stoorvogel and Smaling 1990), a situation nowhere more evident than in the highlands of Tigray Regional State in northern Ethiopia (Gebremedhin 1998; Hagos et al. 1999). Fertilizer use in Tigray is also estimated to be one of the lowest in the country and therefore in the world (Hagos et al. 1999).

Policymakers and development agents have typically followed a two-pronged approach to increasing crop productivity. Various public-led soil conservation programs since the 1970s have focused on conserving soils through physical structures (World Bank 1997; Pagiola 1999; Gebremedhin 1998; Hagos et al. 1999; Admassie 2000). Other efforts to address the problem of nutrient depletion and declining crop productivity focus on promoting yield-enhancing inputs, particularly chemical fertilizers (Ayele and Heidhues 1998; Sasakawa-Global 2000). In 2000 with donor support the Government of Ethiopia initiated a green revolution-type extension program known as Sasakawa Global 2000 (Sasakawa-Global 2000) that emphasizes increasing the use of fertilizer and improved seeds. The program reported some impressive results (Sasakawa-Global 2000, 1996; Demeke et al. 1997), including in Tigray, where fertilizer use increased 13% (Pender et al. 2002).

The literature has proposed several explanations for what many consider the limited adoption of innovations by smallholders. One common explanation focuses on risk averting behavior (Feder et al. 1985; Eicher and Baker 1982). Sandmo (1971), for example, shows that one effect of risk aversion on a profit-maximizing firm is to reduce effort. This implies that producers without perfect insurance will underpurchase inputs, underproduce and underinvest (e.g., Feder et al. 1985; Eicher and Baker 1982).

This view has recently been re-evaluated, because poor rural households do not appear to systematically underproduce given their endowments and markets for agricultural products (Walker and Ryan 1990). Fafchamps (1999) argues that Sandmo's result seems to be counterintuitive; if people are poor and concerned about their survival, the solution is clearly not to underproduce. Finkelshtain and Chalfant (1991) examine the effects of price risk on household behavior and show that households' response to risk is determined by their food security status and their level of dependence on the market. Faced with the same level of risk aversion, in risky environments households that consume some or all of their output produce more than producers who are more involved with the market. Moreover, net-selling households use fewer inputs and produce less under risk, while risk averse net-buying households with severe risk aversion increase their production. Furthermore, Fafchamps (1992) points out that households facing thin and isolated markets often aim for food self-sufficiency. In this case, poor households may apply a risky input if it brings them closer to food self-sufficiency.

While these theoretical developments have encouraged a reconsideration of the conventional wisdom, there is a scarcity of empirical evidence that establishes links between risk aversion and technology adoption. Exceptions are Barrett (1996) who shows that small net-buying households facing food price risk are induced to hyper-exploitation of household labor, and this leads to an inverse relationship between farm size and productivity. Fafchamps and Kurosaki (1997) test the efficiency of insurance markets using data from Pakistan and find that consumption preferences affect crop production choices, suggesting that food security affects household investments. Neither of these papers, however, addresses the role of risk in using purchased farm inputs with a goal of ensuring food security. This chapter joins Chapter 5 by Yesuf and Teklewold in aiming to extend the literature on the effects of risk on fertilizer use.

This chapter explores the link between risk aversion and fertilizer use by examining whether the probability and intensity of fertilizer use is affected by household risk aversion. This is done while controlling for factors such as tenure security, soil/plot characteristics, household and village factors, and possible selection bias. The following section specifies a theoretical framework that incorporates risk into household decision-making. The subsequent two sections discuss data and hypotheses, followed by an econometric model and results. The final part concludes our chapter.

Theoretical Framework

The model used here draws on work by Finkelshtain and Chalfant (1991), Fafchamps (1992; 1999) and Sadoulet and de Janvry (1995). Often studies of behavior under risk assume producers maximize the expected value of utility defined only over income or wealth. Aversion to risk in this single argument is measured by the Arrow-Pratt index of risk aversion (Arrow 1970; Pratt 1964). However, to model households facing risk in other arguments a more general objective function and alternative definition of aversion to risk is needed (Finkelshtain and Chalfant 1991).

We assume a household is engaged in the production of a food crop (m) that it consumes. Output is chosen prior to the realization of prices, while consumption occurs after the harvest when prices are known. The output and input prices are denoted by p and q, wage rate is w, and quantity of inputs (x). The household makes production and consumption decisions to maximize expected utility as given in (6.1).

$$\max_{m,l} E\left[U\left(m,l\right)\right] \tag{6.1}$$

subject to

$$y = pm + wl - qx + T, \tag{6.2}$$

where y denotes full income. Because the optimal consumption plans may be revised *ex post*, the *ex ante* decision involves only leisure (l) and x. Substituting the *ex post* optimal plans for m into U leads to the variable indirect utility function $V(y, p, q, l)$, which is the dual to (6.1) (Epstein 1975, 1980) and can be understood as the result of a two-step optimization process. In the first period the producer chooses how much to work and then how to spend earned and unearned (T) income on consumption.

We only consider the case of output price fluctuations (see Finkelshtain and Chalfant 1991; Sadoulet and de Janvry 1995). Abstracting from the prices of all other commodities that households purchase, the input level x that maximizes expected utility is given by the first order condition:

$$Ev'_y\left[pf'\left(x\right) - q\right] = Ev'_y pf'\left(x\right) - Ev'_y q$$
$$= Ev'_y \overline{p}\left[1 + \sigma_{v'_y}\sigma_p corr\left(v'_y, p\right)\right]f'\left(x\right) - Ev'_y q = 0. \tag{6.3}$$

The impact of risk on production behavior therefore depends on the sign of the correlation between v'_y and p. If p and y are uncorrelated, input use is determined by $\overline{p}f'\left(x\right) = q$ independent of risk, where $\overline{p}$ is the mean price, but if p and y are correlated then input use and supply response are affected by risk. There are two elements affecting the relationship between p and v'_y. First, we have a production effect where an increase in p increases y and decreases the marginal utility of income. Second, there is a consumption effect that reduces income and increases the marginal utility of income.

The net effect is given by the Taylor expansion (Sadoulet and de Janvry 1995):

$$corr\left(v_y', p\right) \approx -\sigma_p^2 \left(R\left(s_p - s_c\right) + \eta s_c\right) \tag{6.4}$$

$$= -\sigma_p^2 \left(Rs_p - s_c\left(R - \eta\right)\right), \tag{6.5}$$

where R is the coefficient of relative risk aversion, η the income elasticity of consumption of food, $s_c = pc/y$ the share of food consumption in total expenditure and $s_p = pf(x)/y$ the share of the risky income in total income. These expressions show that household food self-sufficiency critically determines the impact of uncertainty on production. For net-selling households [*i.e.*, $f(x) > c$; $s_p > s_c$] (6.4) is always negative. Hence, marginal utility of income is negatively correlated with price, implying output under risk is lower than under certainty.

For net buyers (6.4) is negative for low values of R and positive for large values. Net-buying households with mild risk aversion therefore behave as producers and reduce output, though the negative effect on production is lower than for pure producers. Extremely risk averse self-sufficient and net-buying households indeed increase their production in response to risk. This result is seen in (6.5). A positive $R - \eta$ indicates that self-sufficient and net-buying households produce more than pure producers when faced with the same risk aversion.

The intuition behind this result is straightforward. We know that in areas like highland Ethiopia (η) is usually lower than R, because staples constitute a large share of total consumption and have very low-income elasticities. As discussed in the previous chapter, households also tend to be highly risk averse. High transport cost and low agricultural productivity make food markets thin and isolated. Consequently, households are confronted with food prices that are volatile and negatively correlated with their own agricultural outputs. Households therefore try to protect themselves against food price risk. In most cases, this is achieved through self-sufficiency, but it is also true that growing crops whose revenue is positively correlated with prices constitutes insurance. Consequently, more risk averse households seek to insure against consumption price risk by increasing production of staple crops if the covariance condition holds and the direct portfolio effect is not strong enough to induce a reverse behavior. These insights are incorporated into our econometric model.

Study Site, Data Description, and Risk Aversion Estimation

The mainstay of the economy in Tigray, as in the rest of the country, is rainfed mixed crop-livestock agriculture. Agricultural production in the region is highly risky not only because of recurrent drought and adverse weather conditions, but also due to land degradation. Loss in soil fertility is considered a major problem, and soil and water conservation has been ongoing in the region (Geberemedhin 1998; Hagos et al. 1999; Hagos and Holden 2001).

TABLE 6-1   Participation in Food Markets of Farm Households in Tigray Region

Participation status of households	1997	2000
Self-sufficient (non-participating)	2.5	0.5
Net seller	26.7	5.5
Net buyers	70.7	94
Total	100	100

Source: Calculated by the authors.

These conservation efforts are typically complemented with improved nutrient management strategies, such as application of inorganic fertilizer, which lately has become one of the leading strategies for addressing food insecurity.

Northern Ethiopia has relatively limited facilities to support market development (e.g., roads, marketing and storage facilities, etc.). Table 6-1 shows that the overwhelming majority of households in our sample are food deficit (net-buying) households, with only 6.0% of households self-sufficient or net sellers in 2000. We use a cross sectional sample of 400 randomly selected households operating 1,753 plots[1] in 16 villages in the Tigray Regional State of northern Ethiopia. We stratify villages on the basis of agricultural potential as indicated by average annual rainfall and presence of irrigation projects, market access, and population density. To elicit households' subjective risk preferences we use a hypothetical approach similar to that of Binswanger (1980), Sillers (1980), and Wik and Holden (1998). Household heads and their spouses chose between six prospects with different expected yields and levels of risk, with the questions framed to reflect farmers' real production decisions. Respondents were offered X quintals in a good year but no yield in a bad year, and an alternate crop variety that gives Y quintals in a good year and Z quintals in a bad year, with X > Y > Z > 0. They were then asked which crop variety they would prefer to plant. It is assumed that a bad year occurs one out of five years.

Based on responses, unique risk aversion coefficients (R) are derived using constant partial risk aversion (CPRA) utility functions, with the mean serving as a measure of risk aversion. Based on elicited risk preferences, a majority of households exhibits intermediate to extreme risk aversion (see Table 6-2) and only 11% have moderate risk aversion or neutrality. This finding is similar to those in the literature, including those of Yesuf and Teklewold in Chapter 5.

We ran a regression to identify the correlates of risk preferences and find that a Durbin-Wu-Hausman endogeneity test (Davidson and Mackinnon 1993) rejects endogenous risk preferences. Results in Table 6-3 show that most of the estimated coefficients are in line with the literature already discussed. Partial relative risk aversion is negatively correlated with households' consumption expenditures and oxen holdings, which are measures of income and wealth, pointing to decreasing absolute risk aversion (DARA) as found by Binswanger

TABLE 6-2 Risk Aversion Coefficient and Households Classification

a*	b*	c	d	e	f
Good Year	Bad Year	Partial Relative risk Aversion Coefficient**	Bounds of the CPRA Function	Risk Aversion Class	Proportion of the Sample
20	0	0	< = 0	Neutral to preferring	4.7
19.5	2	0.5897	0.5897–0.999	Slight to neutral	1.99
18	4	0.999	0.999–2.4414	Moderate	4.24
16	6	2.4414	2.4414–5	Intermediate	9.5
13	8	5.975	5–14.62	Severe	16.96
9	9	14.62	> = 14.62	Extreme	62.6

*Columns a and b provide the hypothetical expected yield during a good and bad year.
**Constant partial risk aversion coefficients derived using a utility function with constant partial risk aversion.

(1980) and Wik and Holden (1998). Moreover, expected gain is negatively correlated with household risk aversion, suggesting increasing partial risk aversion.

Hypotheses

Many farm households in the study area are poor and net buyers of food, with production mainly for their own consumption. It is therefore likely that households strive to be food self-sufficient in the face of adverse risks and poor market integration (Fafchamps 1999). Though risk and risk aversion may encourage farm households to use fertilizer if it brings them closer to their food security objectives, poor households could use less fertilizer for a variety of reasons, including poorly functioning credit markets. Poor households are also cash constrained at least partly because of underdeveloped or segmented labor markets. They also generally have less productive lands than richer households; relatively poor returns from fertilizer could be a disincentive to fertilizer use.

Given the above, we expect relatively wealthy households (measured by asset endowments) to in general apply more fertilizer, although the specific effect could vary depending on the type and nature of asset holdings. Livestock is expected to increase the use of fertilizer if livestock are important income sources, reducing liquidity constraints and enhancing households' access to formal credit, allowing farmers to intensify production. It is also possible that livestock are themselves important fertilizer sources and manure could substitute for chemical fertilizers. The effect of livestock holding is therefore ambiguous.

Whether poor households use fertilizer depends very much on how binding are credit and cash constraints. In the Ethiopian context farmers obtain credit

TABLE 6-3 Correlates of Risk Preferences

	Dependent Variable: Partial Risk Aversion Coefficient (R)	
Variables	*Coefficient*	*Robust Standard Error*
age of head	0.005	0.012
female-headed	0.016	0.496**
literate head	0.702	0.397*
dependency ratio	−0.381	0.175**
per capita expenditure	−0.0001	0.0001*
female adult labor	−0.347	0.206*
adult male labor	−0.168	0.231
per capita farm holding	−0.094	0.246
per capita oxen holding	−1.734	0.890 **
per capita livestock holding	0.998	0.629
expected return	−2.846	0.055***
cogito2	−1.808	0.513***
cogito3	−1.517	0.615**
enudmy2	−1.515	0.900*
enudmy3	−1.442	0.932
enudmy4	−2.313	0.880***
enudmy5	−1.625	0.932*
cons	47.790	1.630***
/sigma	1.590	0.06

Number of obs = 386
Wald chi2(17) = 6406.55
Log likelihood = −154.32
Prob > chi2 = 0.000
right-censored observations = 257
interval observations = 129

*, **, *** are levels of significance at 10%, 5% and 1% respectively.

either in cash or in-kind from different service providers. Fertilizer is usually available from Bureaus of Agriculture, while cash is provided by microfinance institutions. Poor households' access to food aid and food-for-work programs also provides limited insurance and perhaps cash income, which can be used for risky inputs like fertilizer (Barrett et al. 2001). On the other hand, households' risk exposure is reduced by such programs, which may cause households to focus more on profitability considerations. The relationship between public programs and fertilizer use is therefore ambiguous as well.

Tenure insecurity and the type of land rental arrangements could also affect the use of land enhancing inputs. Tenure insecurity is likely to be less important with respect to the decision to invest in fertilizer than in physical

TABLE 6-4 Summary of Hypotheses and Expected Signs

Variable name	Variable description	Hypothesized expected sign
Risk aversion	A measure of households' risk aversion coefficients	+/–
Resource poverty	Livestock holding	+/–
Resource poverty	Labor endowment	+
Resource poverty	Cash constraint	–
Market access	Distance to market	–/+
Land title: owner operated	Farming own land	+
Land title: rented in	Farming rented-in or transferred-in land	–
Plot distance	Distance from home	–
Land title: presence of trees	Presence of trees on the plot	–/+
Household variables: age	Age of household head	–/+
Household variables: sex	Being female-headed	–
Household variables: education	Education status of head	+
Plot characteristics: soil fertility status	Plots with fertile soils	–
Plot characteristics: degree of erosion	Plots with high level of erosion	–
Plot characteristics: conservation status	Plots with conservation	+

conservation structures, because benefits accrue in the short term. Other incentive problems associated with rented-in plots could, however, reduce the productivity and desirability of applying fertilizer. To test such hypotheses we include duration of tenure and dummies for owner-operated, rented-in, and temporary transfer tenure arrangements in our models; we expect fertilizer to be more often and more intensively used on owner-operated plots than share-cropped or temporarily transferred plots.

We also include distance of plots from homesteads to account for transaction costs involved with managing distant plots and to control for differences in security associated with plot proximity; households are expected to be less likely to use fertilizer on distant plots. On-farm tree planting may also be important. If tenure security is important and trees increase security, plots with planted trees may be selected for fertilizer use. On the other hand, trees may be planted on plots that are not suitable for crop production. The expected sign is, therefore, ambiguous. A summary of our hypotheses is given in Table 6-4.

Econometric Models

Based on the theoretical framework discussed above, we specify the model of fertilizer adoption presented in (6.6):

$$I_i = f(R, Market, Tenure, Crop, P_{cha}, V, h^z) \qquad (6.6)$$

where the dependent variable I_i is fertilizer in kilograms per hectare on plot i and R is the estimated average constant partial risk aversion coefficient. We also include variables related to credit market and safety net access (*Market*), such as access to the formal credit market, food aid and food-for-work. We do not include informal credit, because our data show that households seldom use such credit to purchase fertilizer. Tenure status and duration (number of years operated) and the presence of planted trees as measures of tenure security are included to test whether any tenure variables influence the decision to use fertilizer.

We include type of crop (*Crop*) grown, proxied by the number of plowing days, to adjust for crop choice. Plot characteristics (P_{cha}) include farm level characteristics like soil depth, soil type, susceptibility to erosion, access to irrigation water, whether a plot is conserved or not, and distance from homesteads. Household characteristics (h^z) include age, education and gender of the household head, and the consumer-worker ratio. Household assets include farm size in hectares per capita, livestock holding (TLU, or tropical livestock units, per hectare) and male and female adult laborer per hectare of farmland. Finally, village level variables (V) like agro-climatic factors (rainfall and altitude), market access (distance to major markets), and population density control for village fixed effects.

Unlike some previous literature that use Tobit to model fertilizer demand (e.g., Croppenstedt et al. 1999), we choose an approach similar to the previous chapter and model adoption as a two-stage process, where households first decide whether to use fertilizer and then choose how much to apply on each plot. This allows us to test whether the probability and intensity of input use are determined by different factors. Households also do not randomly decide whether to use fertilizer and indeed self-select whether to participate. As a result, our dependent variable is censored, with many zero values. To account for the two-stage decision process, censoring of the dependent variable, and to control for sample selection bias, we specify the following selection model:

$$\pi = d[x'\beta + u] \qquad (6.7)$$

where the dependent variable is determined by the regressors x and an unobservable error term u. The indicator variable d shows whether the dependent variable is censored or not and is determined through a binary choice model by a vector of conditioning variables z:

$$d = 1[z'\gamma + v \succ 0] \qquad (6.8)$$

where 1[.] denotes an indicator function for applying fertilizer, γ is a vector of unknown coefficients, and v is the unknown error term. If the error terms in equations (6.7) and (6.8) are uncorrelated a separate estimation of the two equations yields consistent results. If this assumption is violated, however, joint estimation of the model is warranted.

The model in (6.7) and (6.8) can be estimated in several ways. Given that $v \sim N(0,1)$ and $E(u|v) = \gamma_1 v$, which simply requires linearity in the population regression of u on v (Wooldridge 2002), (u_1, v_2) is bivariate normal. This model can then be estimated using Heckman's selection model (Heckman 1979). The estimators obtained from this procedure are consistent and $\sqrt{N}$ – asymptotically normal. Identification requires that x is not perfectly correlated with $\lambda(x\delta_2)$.[2] Strictly speaking this can be assured as long as $\lambda(.)$ is a nonlinear function. Thus, we ensure nonlinearity of $\lambda(.)$ by using a linear functional specification in the selection equation while the outcome equation is log-linear.

The Heckman model is sensitive to misspecification, especially violation of assumptions of normality and homoskedasticity of the error terms. In using the Heckman model it is not possible to know whether these assumptions are violated, because it is difficult to test them (Wooldridge 2002). Different models have been proposed that relax the distributional assumptions (Newey 1988; Ahn and Powell 1993). Deaton (1997) follows a procedure similar to Newey (1988) and uses a polynomial form of the predicted probabilities of the binary choice model as an approximation to the inverse Mills ratio. In this chapter we use Deaton's model alongside Heckman's to assure robustness to distributional assumptions.

Using the Deaton model for the intensity equation we estimate robust standard errors to correct for heteroskedasticity (White 1980). We test for heteroskedasticity using the Cook and Weisberg (1983) test and normality using skewness and kurtosis as well as the Shapiro-Wilk and Shapiro-Francia (Gould and Rogers 1991; Gould 1991) normality tests. Tests indicate the presence of heteroskedasticity and violation of the normality assumption. We try linear, log-linear, and (reduced) translog functional forms to eliminate heteroskedasticity. Although it is not possible to eliminate the problem, the log-log functional form yields the lowest χ^2 values. We also test for multicollinearity and find the second and third degree polynomials of the predicted probabilities are highly collinear with a variance inflation factor (VIF) that exceeds 10 (Montgomery and Peck 1992). We therefore only include the third degree polynomial.

Results and Discussion

Descriptive statistics of all variables are reported with standard errors adjusted for stratification and clustering. The average per capita farm holding in the region is 0.3 hectares. Nearly 84% of plots are owner-operated with the remaining being rented-in or temporarily transferred. The average fertilizer use is very low at 9.7 kilograms per hectare. Fertilizer is used on about 38% of 1,507 sample plots.

TABLE 6-5 Description of Variables and Summary Statistics

Variable	Description	Mean	Std. Error*
Fertilizer intensity	fertilizer use in kg per ha	9.71	0.52
Conservation dummy	dummy for conserved land (yes = 1)	0.74	
Plowing days	labor man days used in land preparation per ha	10.77	22.29
Average farm holding	per capita land area (in ha)	0.33	0.02
Trees dummy	presence of planted trees on plot (yes = 1)	0.11	
Soil type dummies	soil type (*hutsa* = 0, *Baekel* = 1, *Mekih* = 2, *Walka* = 3)		
Baekel		0.25	
Mekih		0.26	
Walka		0.24	
Soil depth dummies	soil depth (shallow = 0; medium = 1 and deep = 2)		
Medium		0.39	
Deep		0.38	
Erosion dummies	susceptibility to erosion (none = 0; low = 1; moderate = 2 and high = 3)		
Low		0.29	
Moderate		0.14	
High		0.09	
Plot distance	plot distance from homestead (in minutes)	25.89	2.28
Rainfall variability	coefficient of variation in annual rainfall	29.07	2.74
Altitude dummies	altitude (*hausi kola* = 0, *hausi degua* = 1, *degua* = 2)		
Hausi degua		0.45	
Degua		0.43	
Tenure dummies	type of tenure arrangement (owner-operated = 0 and rented-in = 1, temporary transfer = 2)		
Rented-in		0.15	
Temporarily transferred		0.02	
Tenure duration	tenure holding (in number of years)	11.09	0.44
Irrigation dummy	irrigated plot (yes = 1)	0.06	0.02
Age	age of household head	53.24	1.13
Sex dummy	sex of household head (female = 1)	0.09	0.02
Education dummy	education of household head (literate = 1)	0.39	0.03
Male labor force	number of male adults per ha	1.41	0.07

(continued)

TABLE 6-5 *(Cont.)*

Variable	Description	Mean	Std. Error*
Female labor force	number of female adults per ha	1.33	0.06
Dependency ratio	consumer-worker ratio	3.14	0.11
Average oxen holding	oxen holding per ha	0.18	0.01
Average livestock holding	livestock holding (in tropical livestock units) per ha	0.40	0.04
Credit access dummy	access to formal credit market (yes = 1)	0.32	0.023
Food aid access dummy	access to formal food transfers (yes = 1)	0.79	0.04
Migration dummy	presence of migrant in the household (yes = 1)	0.18	0.02
Population density dummy	population density (dense = 1)	0.61	0.13
Distance to market	distance to a major (*wereda*) market in minutes	138.68	19.89

*Standard errors are adjusted to stratification and cluster effects.

Around 32% of households have access to formal credit from microfinance institutions. The share of migrant household members is relatively small, with only 18% having migrant members supplying remittances. Around 80% of households receive food either from food aid or food-for-work programs, the latter reaching about 57% of households. Most households are located far from major markets with an average walking time to a major (*wereda*) market of about 138 minutes. Close to 61% of households live in villages having a population density of more than 200 persons/km².

Eighty-eight percent of plots are located at altitudes greater than 1,500 meters above sea level, with a fairly uniform distribution of soil types. Most plots are shallow to medium, but only 9% are highly susceptible to erosion. Rainfall is variable in the region with a coefficient of variation of 29% indicating strong inter-village variation. Irrigation plays an insignificant role with only 6% of plots having access to irrigation water. Finally, on average a household has to walk 26 minutes to get to the most distant plot.

Risk Preferences and Probability of Fertilizer Use

Probit results on the probability of fertilizer use are given in Table 6-6. The risk aversion coefficient is highly significant and positive, implying fertilizer use is increasing in risk aversion. This suggests that households who perceive and are presumably trying to manage risks include fertilizer in their risk management portfolios. Livestock increases the probability of using fertilizer, suggesting that livestock relaxes household liquidity constraints, though oxen holding turns out to be statistically insignificant. Not surprisingly, access to formal

TABLE 6-6 Probability of Fertilizer Use

Dependent variable: Binary (0/1)		
Variables	Coefficient	Robust Standard Error
ln(*risk preference*)	0.081	0.035**
Tenure status		
rented-in	−0.041	0.121
transferred	−0.135	0.327
trees	−0.027	0.114
ln(tenure duration)	0.093	0.052*
Household characteristics and asset holding		
ln(age)	−0.072	0.137
female-headed	0.042	0.155
education status	0.083	0.078
ln(female adult labor per ha)	0.149	0.072**
ln(male adult labor per ha)	−0.119	0.068*
ln(dependency ratio)	−0.027	0.083
ln(tropical livestock unit per ha)	0.083	0.043**
ln(oxen adult labor per ha)	0.049	0.047
ln(mean farm size)	−0.207	0.098**
Market access related and village level variables		
access to credit	0.317	0.084***
access to remittances	0.189	0.093**
access to food aid	0.063	0.093
population density	0.103	0.083**
ln(distance to market)	−0.318	0.042***
Plot level variables		
conserved plot	0.381	0.091**
ln(frequency of plowing)	0.261	0.039***
medium soil depth	0.030	0.085
deep soil	0.027	0.096
Baekel	0.032	0.103
Mekih	0.222	0.103**
Walka	0.143	0.115
mid highland	−0.178	0.126
extreme highland	−0.249	0.125**
ln(rain variability)	−0.146	0.120
ln(distance of plot from homestead)	−0.054	0.029**
irrigated plot	0.213	0.159
low susceptibility to erosion	0.069	0.087

(continued)

TABLE 6-6 *(Cont.)*

Dependent variable: Binary (0/1)		
Variables	Coefficient	Robust Standard Error
moderately susceptible to erosion	0.124	0.112
highly susceptible to erosion	−0.037	0.134
_cons	0.453	0.660

Number of obs = 1483
Wald chi2 (31) = 217.32
Prob > chi2 = 0.0000
Log likelihood = −867.14
Pseudo R2 = 0.123

*, **, *** are levels of significance at 10%, 5% and 1% respectively.

credit is positively correlated with the probability of using fertilizer, indicating credit constraints are important. Better access to markets also increases the probability of fertilizer use.

Both rented-in and temporarily transferred plots have lower probabilities of being selected for fertilizer use compared with owner-operated plots, though the results are not statistically significant. The coefficient for trees is also not significant. Length of tenure holding is positive and significant, implying that tenure security increases use of purchased fertilizer. Better tenure security therefore is found to increase short-run farm investments as was also found for more durable investments in Chapter 4 of this volume. Farm size reduces the probability of fertilizer use, and households in densely populated areas have a higher probability of using fertilizer, suggesting population pressure induces intensification. Distance from plots discourages fertilizer use.

Plot productivity variables also influence the probability of fertilizer use. Conserved plots are more likely to be treated than unconserved plots, but the decision to use fertilizer also varies by crop type. Households are most likely to use fertilizer on plots with relatively fertile soils planted with teff, which is typically the highest value crop, followed by wheat, maize, sorghum, and barley. Coefficients for all three soil types (compared with sandy soils) are positive, although only the coefficient for *Mekih* is statistically significant. Finally, the probability of fertilizer use in extreme highlands (*Degua*) is lower than in the middle altitudes (*Hausi Kola*). Rainfall variability is not significant. In terms of household characteristics, households with more adult female labor have higher probabilities of using fertilizer compared to those with more males.

Risk Preferences and Fertilizer Use

The models of fertilizer application are reported in Table 6-7. Both the Heckman and Deaton selection models show a significant selection bias, and

TABLE 6-7 Intensity of Fertilizer Use

Variables	Heckman model		Deaton's model	
	Coefficient	Robust Standard Error	Coefficient	Robust Standard Error
ln(risk preferences)	0.037	0.046	−0.033	0.039
Tenure status				
rented-in	− 0.083	0.166***	0.392	0.172
transferred	0.040	0.358	0.166	0.335
tree	−0.015	0.139	0.095	0.112
ln(tenure duration)	−0.045	0.079	−0.049	0.082
Household characteristics and asset holding				
ln(age)	−0.212	0.181	−0.386	0.167**
female-headed	−0.313	0.279	−0.423	0.228*
education status	0.007	0.095	0.055	0.086
ln(dependency ratio)	0.349	0.286**	0.020	0.095
ln(female adult labor per ha)	0.118	0.132	0.169	0.074**
ln(male adult labor per ha)	0.296	0.310	0.123	0.080
ln(livestock unit per ha)	0.112	0.052**	0.092	0.049**
ln(oxen adult labor per ha)	0.078	0.055	0.027	0.051
ln(mean farm size)	0.252	0.341	0.002	0.125
Market access related and village level variables				
access to credit	0.177	0.106*	0.221	0.107**
access to remittances	0.066	0.113	−0.041	0.097
access to food aid	−0.000	0.109	−0.065	0.096
population density	0.139	0.110	0.091	0.090
ln(distance to market)	− 0.241	0.062***	−0.125	0.051***
Plot level variables				
conserved plot	0.083	0.114	0.025	0.118
ln(frequency of plowing)	0.352	0.036***	0.624	0.054***
medium soil depth	−0.036	0.102	0.041	0.091
deep soil	−0.071	0.117	−0.142	0.093
Baekel	0.071	0.1123	0.189	0.111*
Mekih	0.183	0.129	0.142	0.107
Walka	0.095	0.132	0.195	0.106*
mid highland	−0.040	0.139	0.282	0.150*
extreme highland	−0.043	0.141	0.329	0.148**
ln(rain variability)	−0.493	0.608	−0.094	0.087
ln(distance of plot)	0.004	0.037	0.037	0.036

(continued)

TABLE 6-7 *(Cont.)*

Variables	Heckman model		Deaton's model	
	Coefficient	Robust Standard Error	Coefficient	Robust Standard Error
low susceptibility to erosion	0.010	0.106	0.044	0.086
moderately susceptible to erosion	−0.070	0.136	0.046	0.127
highly susceptible to erosion	−0.076	0.159	−0.044	0.147
_cons	2.116	0.944**	4.156	0.800***
mills	1.774	0.463***	–	–
phat^3	–	–	−1.099	0.377***
	Number of obs = 1177		Number of obs =1484	
	F(34, 1142) =11.67		R-squared = 0.229	
	Prob > F= 0.000		F (34, 350)= 10.79	
	R-squared = 0.234		Prob > F = 0.000	
			Cook-Weisberg test:	
			Chi2(1) = 25.4	
			Prob > chi2= 0.000	

*, **, *** are levels of significance at 10%, 5% and 1% respectively.

tests of normality and homoskedasticity are violated for the Deaton model. Hence, in the Deaton model we estimate robust standard errors using the method of White (1980).

The two models produce similar results. In neither model once households have decided to use fertilizer is the amount of fertilizer used per hectare significantly related to risk preferences. We find instead that households primarily respond to market forces in deciding how much fertilizer to apply. Access to formal credit affects fertilizer applications, as it did the decision to use, underlining the importance of liquidity constraints also in the amount of fertilizer used. Poor access to markets, as measured by distance to major markets, reduces the amount of fertilizer as well as whether it is used. Better access therefore seems to increase fertilizer returns by reducing costs, offering opportunities, and reducing price volatility through better market integration; households in remote areas therefore appear to focus on self-sufficiency and use less fertilizer. Livestock holding has a positive and significant effect on intensity as it did on the decision whether to use fertilizer and oxen holding is again not statistically significant. Finally, the intensity of fertilizer use varies strongly with crop type and more is applied to higher value crops.

Conclusion

This chapter examines how the probability and level of fertilizer use are affected by risk aversion while controlling for a host of biophysical, household, institutional, and village factors. Results show that risk aversion, as measured by the average constant partial risk aversion coefficient, positively affects households' decisions to use fertilizer, but it has no effect on applications per hectare. One possibility is that risk aversion increases adoption as long as fertilizer contributes to increased food self-sufficiency. These findings are almost the opposite of those in the previous chapter, which used similar methods and found that risk aversion reduces both fertilizer adoption and use intensity.

Also in contrast to findings of Yesuf and Teklewold in Chapter 5, the decision to use fertilizer, as well as applications per hectare, are also influenced by some important non-risk factors. Wealth measured by livestock holdings increases the probability and amount of fertilizer use as does access to formal credit. Poor market access reduces the probability and intensity of fertilizer use, likely due to depressed marketing opportunities, increased price volatility, higher transaction costs in accessing fertilizer markets, and less reliable fertilizer supply. Price volatility due to segmented markets is very important in Ethiopia due to poor infrastructure, which causes prices of agricultural commodities to fall drastically during good years and skyrocket during droughts.

These results suggest that even in extremely low-income environments like Tigray risk aversion does not hamper efforts to increase the application of commercial fertilizers. Instead, poverty and missing or segmented markets inhibit adoption. While the role of fertilizer in food security merits further scrutiny, our evidence points to the need to focus on the development of market infrastructure and credit institutions if the existing low level of fertilizer application in Ethiopia is to be increased. Aversion to risk does not appear to be a constraining factor.

Acknowledgments

We gratefully acknowledge the Norwegian Research Council and Policies for Sustainable Land Management in the Highlands of Tigray, an IFPRI/ILRI/ Mekelle University Research project, for supporting our fieldwork. We are also grateful to participants at an International Seminar at Wageningen University in July 2002 and to John Pender and Mette Wik for their valuable comments. The usual disclaimer applies.

Notes

1   We exclude rented-out plots, leaving us with a sample of 1,507.
2   The λ represents what is usually known as Heckman's lambda, which is given
as: $\lambda = \dfrac{\phi(z_i'v)}{\Phi(z_i'\gamma)}$ where $\phi(.)$ and $\Phi(.)$ are the corresponding PDF and CDF.

References

Admassie, Y. 2000. *Twenty Years to Nowhere: Property Rights, Land Management, and Conservation in Ethiopia.* Lawrenceville, NJ: Red Sea Press.
Ahn, H., and J. L. Powell. 1993. Semi-parametric Estimation of Censored Regression Models with Nonparametric Selection Mechanism. *Journal of Econometrics* 58: 3–29.
Arrow, K. J. 1970. Aspects of the Theory of Risk Bearing. In K. J. Arrow (ed) *Essays in the Theory of Risk Bearing,* Amsterdam: North-Holland Co.
Ayele, G., and F. Heidhues. 1998. Analysis of Innovation, Dissemination, and Adoption of Vertisol Technology: Some Empirical Evidences from the Highlands of Ethiopia. Paper presented at the Conference on Soil Fertility Management. April 21–23, 1998, Addis Ababa, Ethiopia.
Barrett, C. 1996. On Price Risk and the Inverse Farm Size-Productivity Relationship. *Journal of Development Economics* 51: 193–215.
Barrett, C., S. Holden, and D. C. Clay. 2001. Can Food-for-Work Reduce Vulnerability? Final Draft. In *Insurance Against Poverty,* edited by S. Dercon. World Institute for Development Economics Research. United Nations University.
Binswanger, H. 1980. Attitudes toward Risk: Experimental Measurement of Evidence in Rural India. *American Journal Agricultural Economics* 62 (3): 395–407.
Cook, R. D., and S. Weisberg. 1983. Diagnostics for Heteroskedasticity in Regressions. *Biometrica* 70: 1–10.
Croppendstedt, A., M. Demeke, and M. Meshi. 1999. An Empirical Analysis of Demand for Fertilizer in Ethiopia. *Ethiopian Journal of Agricultural Economics* 3(1): 1–39.
Davidson, R., and J. G. Mackinnon. 1993. *Estimation and Inference in Econometrics.* New York: Oxford University Press.
Deaton, A. 1997. *The Analysis of Household Surveys: A Microeconometric Approach to Development Policy.* Published for the World Bank. Baltimore and London: The Johns Hopkins University Press.
Demeke, M., A. Seid, and T.S. Jayne. 1997. Promoting Fertilizer Use in Ethiopia: The Implications of Improving Grain Market Performance, Input Market Efficiency, and Farm Management. Grain Market Research Project Working Paper 5. Addis Ababa, Ethiopia: Ministry of Economic Development and Cooperation.
Eicher, C., and D. Baker. 1982. Research on Agricultural Development in Sub-Saharan Africa: A Critical Survey. MSU International Development Paper No. 1. East Lansing MI: Michigan State University.
Epstein, L. 1975. A Disaggregated Analysis of Consumption Choice under Uncertainty. *Econometrica* 43: 877–92.
———. 1980. Decision Making and the Temporal Resolution of Uncertainty. *International Economic Review* 21(2): 269–83.

Fafchamps, M. 1992. Cash Crop Production, Food Price Volatility, and Rural Market Integration in the Third World. *American Journal Agricultural Economics* 74(1): 90–99.

———. 1999. Rural Poverty, Risk, and Development. Economic and Social Development Paper No. 144. Food and Agriculture Organization (FAO).

Fafchamps, M., and T. Kurosaki. 1997. Insurance Market Efficiency and Crop Choices in Pakistan. Department of Economics, Stanford University. Mimeo.

Feder, G., R. E. Just, and D. Zilberman. 1985. Adoption of Agricultural Innovations in Developing Countries: A Survey. *Economic Development and Cultural Change* 33: 255–98.

Finkelshtain, I., and J. A. Chalfant. 1991. Marketed Surplus under Risk: Do Peasants Agree with Sandmo? *American Journal Agricultural Economics* 73(3): 557–67.

Gebremedhin, B. 1998. *The Economics of Soil Conservation Investments in Tigray Region of Ethiopia*. Ph.D. thesis, Department of Agricultural Economics, Michigan State University, East Lansing, MI.

Gould, W. W. 1991. Sg3: Skewness and Kurtosis Tests of Normality. *Stata Technical Bulletin* 1: 20–21.

Gould W. W., and W. H. Rogers. 1991. Sg3: Summary of Tests for Normality. *Stata Technical Bulletin* 3: 20–23.

Heckman, J. 1979. Sample Selection as a Specification Error. *Econometrica* 47: 153–61.

Hagos, F., J. Pender, and N. Gebreselassie. 1999. Land Degradation and Strategies for Sustainable Land Management in the Ethiopian Highlands: Tigray Region. Socioeconomic and Policy Research Working Paper No. 25. Nairobi, Kenya: International Livestock Research Institute.

Hagos, F., and S. Holden. 2001. Incentives for Conservation in Tigray, Ethiopia: Findings from a Household Survey. Department of Economics and Social Sciences, Agricultural University of Norway. Unpublished.

Montgomery C. D., and E. A. Peck. 1992. *Introduction to Linear Regression Analysis*. Second Edition. John Wiley and Sons, Inc.

Newey, W. K. 1988. Two-Step Series Estimation of Sample Selection Models. Princeton, NJ: Department of Economics, Princeton University. Manuscript.

Pagiola, S. 1999. Global Environmental Benefits of Land Degradation Control on Agricultural Land. Washington, DC: The World Bank. Mimeo.

Pender, J., B. Gebremedhin, and M. Haile. 2002. Livelihood Strategies and Land Management in the Highlands of Tigray. Paper presented in a conference on Policies for Sustainable Land Management in the East African Highlands. Addis Ababa, Ethiopia. Mimeo.

Pratt, J. W. 1964. Risk Aversion in the Small and in the Large. *Econometrica* 32: 122–36.

Sadoulet, E., and A. de Janvry. 1995. *Quantitative Development Policy Analysis*. Baltimore and London: The Johns Hopkins University Press.

Sandmo, A. 1971. On the Theory of the Competitive Firm under Price Uncertainty. *American Economic Review* 61: 65–73.

Sasakawa-Global. 1996. *Agriculture Project in Ethiopia. Annual Report, Crop Season 1995*. P.O. Box 12771, Addis Ababa, Ethiopia.

———. 2000. *Agriculture Project in Ethiopia. Annual Report, Crop Season 1995*. P.O. Box 12771, Addis Ababa, Ethiopia.

Sillers, D. A. 1980. *Measuring Risk Preferences of Rice Households in Nueva Ecija, Philippines: An Experimental Approach*. Ph.D. thesis, Yale University.

Stoorvogel, J. J., and E. M. A. Smaling. 1990. *Assessment of Soil Nutrient Depletion in Sub-Saharan Africa: 1983–2000.* Volume 1: Main Report. Report 28. Wageningen, the Netherlands: The Winand Staring Center.

Walker T. S., and J. Ryan. 1990. *Village and Household Economics in India's Semi-arid Tropics.* Baltimore and London: The Johns Hopkins University Press.

White, H. 1980. A Heteroskedasticity-Consistent Covariance Matrix Estimator and a Direct Test for Heteroskedasticity. *Econometrica* 48: 817–30.

Wik, M., and S. Holden. 1998. Experimental Studies of Peasants' Attitudes toward Risk in Northern Zambia. Discussion Paper No. D-14. Ås, Norway: Department of Economics and Social Sciences, Agricultural University of Norway.

Wooldridge, M. J. 2002. *Econometric Analysis of Cross Section and Panel Data.* Cambridge, MA: MIT Press.

World Bank. 1997. *Soil Fertility Initiative for Sub-Saharan Africa.* Agriculture Group Two, Africa Region. Washington, DC: The World Bank.

CHAPTER 7

Crop Biodiversity and the Management of Production Risk on Degraded Lands: Some Evidence from the Highlands of Ethiopia

SALVATORE DI FALCO AND JEAN-PAUL CHAVAS

A s has been discussed in the previous chapters, difficult climatic conditions and lack of soil nutrients pose important challenges for farm households in Ethiopia (Dercon 2004). During the last fifty years, a large number of severe droughts occurred and crop production in most areas "never topped subsistence levels" (REST 1995, *137*). Dercon (2005) reports that 78% of rural households are seriously affected by harvest failure due to weather variability over time and space. Land degradation exacerbates these risks, and Ethiopia has one of the highest rates of soil nutrient depletion in sub-Saharan Africa (Grepperud 1996; FAO 2001).[1]

Managing risk in such environments and buffering its adverse effects on welfare is therefore very important for agricultural households (Bromley and Chavas 1989; Paxson 1992; Fafchamps 1992; Deaton 1992; Fafchamps and Pender 1997; Fafchamps et al. 1998). *Ex ante* management is particularly important (Just and Candler 1985; Fafchamps 1992; Chavas and Holt 1996; Dercon 1996; Smale et al. 1998), because *ex post* insurance mechanisms often function poorly, due to credit constraints, information asymmetries, and commitment failures (Deaton 1989; Fafchamps 1992; Kurosaki and Fafchamps 2002); when present, safety nets often provide only limited support (Dercon and Krishnan 2000; Dercon 2004). Unfortunately, *ex ante* management options are limited because of missing markets, agro-ecological conditions, and land quality heterogeneity.

In isolated dry environments farmers rely heavily on genetic resources, and Ethiopia is a recognized global center of genetic diversity for several cereals (Vavilov 1949; Harlan 1992). Crops are produced mainly from farmers' varieties, often called "landraces," which are characterized by high morphogenetic variation that has been shaped by natural processes and selection practices

99

over time (Smale et al. 2001). Local germplasm can be preferred to improved varieties, because of its greater tolerance to severe biotic and abiotic stresses[2] (Tesemma and Bechere 1998). Given that crops respond differently to weather, maintaining diversity ensures that "whatever the environmental conditions, there will be plants of given functional types that thrive under those conditions" (Heal 2000, 4).

This chapter investigates how crop biodiversity can contribute to farm productivity and assist with the management of risk. To investigate the potential role of crop biodiversity on risk management we adopt a stochastic production function similar to Antle (1983) that analyzes the impacts of biodiversity on mean, variance, and skewness of production, where skewness captures the effects of biodiversity on downside risk exposure (e.g., the probability of crop failure).

Under risk aversion farm households suffer welfare losses when they experience unexpected income fluctuations. Such fluctuations are measured by the variance (Just and Pope 1979), but the variance does not distinguish between unexpected bad and good events. It therefore seems important to go beyond variance and introduce the chance of crop failure—skewness—into the analysis. Empirical evidence also suggests that most decisionmakers exhibit decreasing absolute risk aversion (e.g., Binswanger 1981; Chavas and Holt 1996), which implies "downside risk aversion" (Menezes et al. 1980; Antle 1983). Farmers therefore have incentives to grow cultivars and varieties that positively affect the distribution of returns, thus reducing their exposure to downside risk like severe droughts leading to crop failure. This chapter investigates the determinants of mean output, variance, and skewness using data on barley and cereals production from the highlands of northern Ethiopia. Dependent variables analyzed are the first three moments of the production function.

Background and Data Information

Our study area is in the Tigray Regional State where most of the population depends on mixed crop and livestock farming and oxen are used for land preparation and threshing. Average annual rainfall is 652 mm, but spatial variability is very high (four times more than the rest of country). For instance, we can observe about 1000 mm in the west and 400 mm in the east. Cereal production is very important (85% of the cultivated land), and 76% of sample farms grow more than one cereal (Pender et al. 1999). In 98 villages surveyed, teff is grown in 83, barley in 78, wheat in 51, millet in 47, and sorghum in 38.

Ethiopia is a center of origin for teff and a center of diversity for barley and wheat (Asfaw 2000). For instance, in crop genetic conservation, Ethiopian barley has been identified as a priority crop since the 1920s, with extensive germplasm collections deposited in gene banks throughout the world (Negassa 1985; Asfaw 2000). While some morphotypes are widely distributed, others are used in more isolated areas (Asfaw 2000). Ethiopian barley is an important source of genes for resistance and protein and many lines are used as donors for commercial varieties in North America and Europe (Qualset and Moseman 1966; Qualset 1975; Negassa 1985; Alemayehu 1995; Asfaw 2000).

Table 7-1 reports variables and definitions, and Table 7-2 presents descriptive statistics. Intra-species diversity within barley (barley diversity) and interspecies diversity across all cereals (cereal diversity) are investigated. As metrics for barley diversity, we choose the Margalef index, defined as [(number of barley varieties)/ln(barley area) – 1]. Cereal diversity is captured by the Shannon index defined as $H = -\Sigma_i p_i \cdot \log(p_i)$, where p_i is the land share under each cereal. Both indices are widely used and capture different aspects of diversity. The former captures richness in barley varieties. The latter refers to the amount of diversity found in a geographic area.

In addition to crop genetic diversity, explanatory variables include conventional inputs (land, labor, animal, urea), environmental and soil conditions (erosion and water logging, slope, fertility, and altitude), and managerial variables (years of experience, number of barley plots). Land, labor, and animals are the most important conventional inputs. The average input of labor is 70 person days and of animals 40 oxen days. Regarding the environmental and soil conditions of farms in the sample, on average 5% of operated plots are affected by severe erosion and water logging and 7% are located on steep slopes. The most important problem is low soil productivity: 37% of plots are infertile. An average of over nine years was spent cropping (maximum of 14 years and a minimum of one) about three plots per household. As of 1998 only 15% of

TABLE 7-1 Variable Definitions

Urea	Fertilizer use (in logs)
Land	Land for barley (in logs)
Labor	Labor in person days (in logs)
Animal	Animal in oxen days (in logs)
Barley biodiversity	Margalef index for biodiversity [(number of barley varieties)/ln(barley area) – 1]
Cereal biodiversity	Shannon index for biodiversity $H = -\Sigma_i p_i \cdot \log(p_i)$
Biodiversity × land fertility	Interaction between biodiversity and fertility
Altitude	Household altitude
Steep slope	Share of land on steep slopes
Severe erosion	Share of land affected by severe erosion and water logging
Fertility	Share of land on medium/high fertility
Land in other crops	Share of land allocated to other crops (in logs)
Experience	Number of years cropping the plots

TABLE 7-2 Descriptive Statistics

	Mean	Std. Dev.	Minimum	Maximum
Urea	1.87598	1.55254	0	4.605
Land	7.93202	0.796989	5.594	9.458
Labor	4.19669	0.612621	1.791	5.676
Animal	3.58615	0.55668	1.098	5.41
Barley Biodiversity	0.167732	0.055073	0.118	0.407
Cereals Biodiversity	0.33456	0.15	0.10	0.87
Altitude	2341.45	305.052	1521	2988
Steep slope	0.083127	0.209678	0	1
Severe Erosion	0.0512646	0.142385	0	0.871
Fertility	0.609767	0.285206	0	1
Land in other crops	8.24783	1.23499	1.203	9.675
Experience	9.20635	2.28426	1	14

household heads had formal schooling (only 6% had over two years), and 7% participated in literacy programs (Pender and Gebremedhin 2004).

Method

Let y represent output and x inputs under risk. The production technology is represented by the stochastic production function $y = g(x, v)$, where v is a vector of random variables reflecting uncontrollable factors affecting output. To assess the probability distribution of $g(x, v)$ we follow Antle (1983) and use a moment-based approach. Consider the following econometric specification for $g(x, v)$:

$$g(x, v) = f_1(x, \beta_1) + [f_2(x, \beta_2) - (f_3(x, \beta_3)/k)^{2/3}]^{1/2} e_2(v)$$

$$+ [f_3(x, \beta_3)/k]^{1/3} e_3(v), \tag{7.1}$$

where $f_2(x, \beta_2) > 0$, $(f_2)^3 \geq (f_3/k)^2$, and the random variables $e_2(v)$ and $e_3(v)$ are independently distributed and satisfy $E[e_2(v)] = E[e_3(v)] = 0$, $E[e_2(v)^2] = E[e_3(v)^2] = 1$, $E[e_2(v)^3] = 0$, and $E[e_3(v)^3] = k > 0$. This means that the random variables $e_2(v)$ and $e_3(v)$ are normalized with mean zero and variance 1. In addition, $e_2(v)$ has zero skewness $(E[e_2(v)^3] = 0)$, but the random variable $e_3(v)$ is asymmetrically distributed and has positive skewness $(E[e_3(v)^3] = k > 0)$. It follows from (7.1) that

$$E[g(x, v)] = f_1(x, \beta_1), \tag{7.2a}$$

$$E[(g(x, v) - f_1(x, \beta_1))^2] = f_2(x, \beta_2), \tag{7.2b}$$

$$E[(g(x, v) - f_1(x, \beta_1))^3] = f_3(x, \beta_3).$$ (7.2c)

The specification (7.1) therefore provides a convenient representation of the first three central moments of the distribution of $g(x, v)$.[3] Indeed, from (7.2a) the first moment (the mean) is given by $f_1(x, \beta_1)$. From (7.2b) the second central moment (the variance) is given by $f_2(x, \beta_2) > 0$, and from (7.2c) the third moment (measuring skewness) is given by $f_3(x, \beta_3)$. In addition, if we treat the distribution of $e_2(v)$ and $e_3(v)$ as given, the three moments $f_1(x, \beta_1)$, $f_2(x, \beta_2)$, and $f_3(x, \beta_3)$ are sufficient statistics for the distribution of $g(x, v)$ in Equations 7.2a, 7.2b and 7.2c. As such, the specification (7.1) expands on previous studies of crop genetic diversity (Smale et al. 1998; Widawsky and Rozelle 1998; Di Falco and Perrings 2005).

In general, one expects mean output $f_1(x, \beta_1)$ to exhibit positive and decreasing marginal productivity with respect to x, $\partial f_1/\partial x > 0$ and $\partial^2 f_1/\partial x^2$ being a negative definite matrix. However, the effects of inputs x on the variance and skewness of output are empirical issues. For example, from (7.2b) the ith input can be variance increasing, neutral, or decreasing. Similarly, from (7.2c) the ith input can increase or decrease downside risk exposure when $\partial f_3/\partial x_i > 0$ (< 0). Of special interest are the effects of genetic diversity on the variance and skewness of production.

Equation (7.1) can be interpreted as a standard regression model where the dependent variable is cereal, or barley production $y = g(x, v)$, $f_1(x, \beta_1)$ is the regression line representing mean effects, and $\{[f_2(x, \beta_2) - (f_3(x, \beta_3)/k)^{2/3}]^{1/2}$ $e_2(v) + [f_3(x, \beta_3)/k]^{1/3} e_3(v)\}$ is an error term with mean zero, variance $f_2(x, \beta_2)$, and skewness $f_3(x, \beta_3)$. The error term exhibits possible heteroskedasticity (given by $f_2(x, \beta_2)$) and skewness (given by $f_3(x, \beta_3)$). Let (i, j) denote the ith farm household in the jth location. Assume that $f_1(x_{ij}, \beta_1) = \beta_{0j} + x_{ij} \beta_1$, where β_{0j} represents unobserved factors influencing productivity in the jth location. It follows that output of the ith household in the jth location can be written as:

$$y_{ij} = \beta_{0j} + x_{ij} \beta_1 + u_{1ij},$$ (7.3a)

where u_{1ij} is an error term with mean zero. Denote by $\bar{y}_j$ and $\bar{x}_j$ the mean of y_{ij} and x_{ij} in the jth location. Then, equation (7.3a) can be written as

$$y_{ij} - \bar{y}_j = (x_{ij} - \bar{x}_j) \beta_1 + u_{1ij}.$$ (7.3a')

Equation (7.3a') involves variables measured as deviations from their location means. In the spirit of Barrett et al. (2004)[4] the specification (7.3a') corrects for the effects of location-specific factors and controls for unobservable heterogeneity related to institutional factors and spatial agro-climatic conditions. After controlling for such effects, the error term u_{1ij} can be interpreted as reflecting production uncertainty facing the ith household. Following Antle (1983) we first estimate (7.3a'), yielding β_1^e, a consistent estimator of β_1 and the associated error term $u_1^e = y - f(x, \beta_1^e)$. We then apply feasible generalized least-squares to (7.3b), generating consistent estimators of the parameters β_k, $k = 2, 3$ (Antle 1983).

$$(u_1^e)^k = f_k(x, \beta_k) + u_k, k = 2, 3. \tag{7.3b}$$

The variance of u_1 in (7.3a) is $f_2(x, \beta_2)$, and the variance of u_k in (7.3b) is $[f_{2k}(x, \beta_{2k}^e) - f_k(x, \beta_k^e)^2]$, $i = 2, 3$ (see Antle 1983). Equations (7.3a) and (7.3b) exhibit heteroskedasticity and a weighted regression approach is therefore used to capture efficiency gains. The effects of farm-specific soil quality, fragmentation, and altitude are represented by the x's in equations (7.3). In addition, each farmer's ability is captured by his or her experience.

We examine whether the model may be subject to endogeneity bias. This would occur if some of the explanatory variables were correlated with the error term. For example, if barley or cereal biodiversity were correlated with the error term, the least-squares estimate of the effects of variety richness on the mean, variance and skewness of output would be biased. We identify a set of suitable instruments[5] following both theory and existing literature using the same database (Pender et al. 1999). The instruments are walking distance from input supplier and lagged values for fragmentation and biodiversity. We test the overidentification restrictions using a Hansen test and—as reported at the bottom of Table 7-3—the instruments pass the test. Instrumental variable estimation is then used to test for endogeneity using a Durbin-Wu-Hausman test (see Davidson and MacKinnon 1993; Wooldridge 2002). We find a Wu-Hausman F test statistic of 2.1 (P value: 0.33) for the barley biodiversity index, suggesting no evidence of endogeneity bias. When the same test is implemented for cereals we find $F = 4$ (P value 0.046). We therefore reject the null of exogeneity and use an IV approach.

Econometric Results

The mean, variance and skewness function estimates are reported in tables 7-3 and 7-4. A linear-log specification is used for the mean and variance, and skewness functions are assumed to be linear. Results are similar for cereals and barley. In the mean function, conventional inputs have positive marginal effects, with labor and land statistically significant; oxen and urea are statistically significant only in the cereals model. We assume that biodiversity enters as a linear function and in interaction with land degradation. We find that only biodiversity is statistically significant in both mean functions, indicating that increasing biodiversity has a positive effect on production. Evaluated at sample means, the elasticity of barley production with respect to biodiversity is 0.55, while the elasticity of cereal output with respect to biodiversity is 0.16.

TABLE 7-3 Barley and Cereal Production Function

	Barley *Fixed Effects; N = 190* *(A)*	*Cereals* *Fixed Effects and IV; N = 367* *(B)*
Urea	7.06 (11.8)	0.13*** (0.004)
Land	120.06*** (25.11)	51.05*** (39.62)
Labor	59.44[a] (42.11)	37.68 (55.65)
Animal	41.99 (52.33)	133.54*** (40.8)
Biodiversity	1280* (680)	1217*** (324)
Biodiversity × fertility	−527.59 (934.79)	−828.27** (398.9)
Altitude	0.11 (0.13)	0.43** (0.21)
Steep slope	10.19 (80.16)	103[a] (79.7)
Severe erosion	113.87 (113.92)	21.54 (34.12)
Fertility	50.56 (172.09)	350.9** (149.32)
Land in other crops	−20.48[a] (13.9)	− −
Experience	10.83[a] (8.09)	17.88** (8.66)

GLS estimates. R-square = 0.615 and 0.47 respectively; Breusch-Pagan chi-squared = 93.1525. Significance levels are denoted by one asterisk (*) at the 10% level, two asterisks (**) at the 5% level, three asterisks (***) at the 1% level, and one ([a]) at the 10% with two-sided test. Constants not reported.

Results for the variance function are shown in the second and fourth columns of Table 7-4. Diversity is statistically significant in both the linear form and in interaction with land degradation, with biodiversity increasing variance of output. The negative sign on the interaction term, however, implies that the effect of variety richness is sensitive to the share of fertile land. The marginal impact of biodiversity is found to be positive when fertility is at the sample mean, but it becomes negative when the share of fertile land is above 80% of plots.

Table 7-4 suggests that biodiversity and fertility should be considered risk-increasing, with a negative effect on the welfare of risk averse farmers.

TABLE 7-4 Variance and Skewness Functions—Fixed Effects Models

Variables	Variance Function $f_2(x, \beta_2)$	Skewness Function $f_3(x, \beta_3)$	Variance Function $f_2(x, \beta_2)$	Skewness Function $f_3(x, \beta_3)$
	Barley biodiversity		*Cereals biodiversity*	
Urea	1778.51[a]	560.023	5.17***	2.01*
	(1146.87)	(797.289)	(1.29)	(1.17)
Land	7207.1***	1034.58	14501.4	−13713
	(2471.16)	(1696.06)	(1279)	(10976)
Labor	−550.978	−2042.38	−34690.65**	−18787.2
	(3798.24)	(2847.58)	(17109.28)	(15418)
Animal	3940.7	4368.27	21947.21*	9960.703
	(4511.14)	(3542.62)	(12559.36)	(11318.55)
Biodiversity	261054***	118055***	299366.1***	259331.9***
	(100809)	(46520)	(99826.6)	(89964.18)
Biodiversity × fertility	−180897*	−121519**	−182239.3[a]	−200352*
	(107488)	(63600.9)	(122622.8)	(110508.2)
Altitude	−2.3271	−3.8209	49.61	−19.06
	(13.9689)	(8.9078)	(59.4)	(53.69)
Steep slope	−2665.81	−2240.78	−43316.9*	−27220.67
	(6711.65)	(5427.8)	(24542)	(22118.16)
Severe erosion	9409.68	−640.363	−9197.4	15741.25
	(9683.41)	(7688.28)	(10506)	(9468.888)
Fertility	34272.9*	20248.2*	69817	56503.25[a]
	(18744.2)	(11674.4)	(46231.7)	(41665.17)
Land in other crops	−1336.07	−529.909	–	–
	(912.748)	(947.215)		
Experience	−121.475	140.51	2231.07	2290.156
	(618.629)	(546.949)	(2667.87)	(2404.299)
Constants not reported.				

However, as discussed earlier, variance does not distinguish between upside and downside risk. Indeed, under very difficult agro-climatic conditions, farmers may be especially averse to "downside risk," which variance fails to capture. Regression results for the skewness function are shown in the third and fifth columns of Table 7-4. Biodiversity is positively and strongly related to skewness of output, which implies that increasing the number of varieties or crops hedges against the risk of crop failure. More varietal diversification, therefore, reduces exposure to downside risk and helps insure food production will not fall below critical levels. Households producing barley at higher altitudes are less exposed to risk, perhaps because cooler temperatures reduce yield variability. Labor has a positive effect on variability, and more oxen may reduce risk. These estimates are not significant.

Simulation

To analyze the economic and welfare implications of our econometric results, we present simulations of the effects of crop biodiversity on risk (See Di Falco and Chavas 2009). The simulations assume that risk preferences exhibit constant relative risk aversion, with a coefficient of relative risk aversion equal to 2.5.[6] Consider the following maximization problem

$$\text{Max }\{EU(c_1, N(x) + p_1 [g(x, v) - c_1])\}, \tag{7.4}$$

where E is the expectation operator. Under non-satiation in c_2 (where $\partial U/\partial c_2 > 0$) the choice of x in (7.4) can be written in terms of the "certainty equivalent" CE satisfying

$$U(c_1, CE - p_1 c_1) = EU(c_1, N(x) + p_1 [g(x, v) - c_1]), \tag{7.5}$$

with $\pi = N(x) + p_1 [g(x, v)]$ and—following Pratt (1964)—equation (7.5) can be expressed as

$$U(c_1, E(\pi) - R - p_1 c_1) = EU(c_1, \pi - p_1 c_1), \tag{7.5'}$$

where $E(\pi)$ is expected income and R is a risk premium measuring the cost of private risk. The risk premium R in (7.5') measures willingness to pay for an insurance scheme that would replace the random variable π by its mean. Combining (7.5) and (7.5') implies that the certainty equivalent can be decomposed into two additive parts:

$$CE = E(\pi) - R. \tag{7.6}$$

By definition risk aversion corresponds to a positive risk premium ($R > 0$). In general, under risk aversion risk exposure lowers the certainty equivalent and makes the decisionmaker worse off. As shown by Pratt (1964), risk aversion can be assessed "locally" using the Arrow-Pratt risk aversion coefficient $r_2 \equiv -(\partial^2 U/\partial \pi^2)/(\partial U/\partial \pi)$. With $\partial U/\partial \pi > 0$, risk aversion corresponds to $R > 0$, $\partial^2 U/\partial \pi^2 < 0$ and $r_2 > 0$. Pratt (1964) defined decreasing absolute risk aversion (DARA) where increasing mean income reduces R, implying that expected income behaves as a substitute for "insurance." Pratt (1964) showed that under DARA $\partial r_2/\partial \pi < 0$, which is of interest because empirical evidence suggests that most decisionmakers exhibit risk aversion and DARA risk preferences (e.g., Binswanger 1981; Chavas and Holt 1996).

Crop biodiversity is one of the inputs in x. Since $\partial r_2/\partial \pi = -(\partial^3 U/\partial \pi^3)/(\partial U/\partial \pi) + r_2^2$, DARA implies $\partial^3 U/\partial \pi^3 > 0$, corresponding to aversion to unfavorable "downside risk" (see Menezes et al. 1980). Under "downside risk aversion"[7] farmers avoid the risk of crop failure (Menezes et al. 1980; Antle

1983), which brings up the question of how maintaining biodiversity affects skewness of the distribution. As shown by Pratt (1964), taking a Taylor series approximation on both sides of equation (7.5') evaluated at point $E(\pi)$ gives

$$U(c_1, E(\pi) - p_1 c_1) - (\partial U/\partial \pi)R \approx U(c_1, E(\pi) - p_1 c_1)$$
$$+ \tfrac{1}{2}(\partial^2 U/\partial \pi^2)E[\pi - E(\pi)]^2 + 1/6(\partial^3 U/\partial \pi^3)E[\pi - E(\pi)]^3.$$

This yields the following approximation to the risk premium:

$$R_a = 1/2\, r_2\, M_2 + 1/6\, r_3\, M_3, \qquad (7.6')$$

where $M_k = E[\pi - E(\pi)]^k$ is the kth central moment of the distribution of profit, $r_2 = -(\partial^2 U/\partial \pi^2)/(\partial U/\partial \pi)$ is the Arrow-Pratt coefficient of absolute risk aversion and $r_3 = -(\partial^3 U/\partial \pi^3)/(\partial U/\partial \pi)$, all evaluated at $E(\pi)$. Equation (7.6') can also be written as:

$$R_a = R_{a2} + R_{a3}, \qquad (7.6'')$$

where $R_{a2} \equiv 1/2\, r_2\, M_2$ and $R_{a3} \equiv 1/6\, r_3\, M_3$. Equations (7.6') and (7.6'') decompose the risk premium R_a into two additive parts: $R_{a2} \equiv 1/2\, r_2\, M_2$ reflecting the effect of the variance M_2 and $R_{a3} \equiv 1/6\, r_3\, M_3$ indicating the impact of skewness M_3 on the cost of risk. When $M_3 = 0$, equation (7.6) reduces to the standard Arrow-Pratt approximation, and the approximate risk premium R_a is (locally) proportional to the variance of production returns M_2, with $r_2/2$ as the coefficient of proportionality (Pratt 1964). It gives the intuitive result that under risk aversion (when $\partial^2 U/\partial \pi^2 < 0$ and $r_2 > 0$) any increase in variance tends to increase the private cost of risk bearing.

Equation (7.6) extends this result to show how the third moment M_3 affects the risk premium, and we find that $\partial R_a/\partial M_3 \approx 1/6\, r_3$ (i.e., the risk premium is decreasing in skewness when $\partial^3 U/\partial \pi^3 > 0$ and $r_3 < 0$). This occurs because higher skewness is a decrease in downside-risk, reducing the cost. This raises at least two questions related to risk management. How does risk affect the incentive to use crop biodiversity as a means of reducing the cost of risk? And what is the relative importance of the variance effect versus skewness in the valuation of the cost of private risk? Answering these questions requires evaluating the risk premium R. As just discussed, this can be done using the risk premium R_a given in equations (7.6') and (7.6'').

We are also interested in exploring farmers' certainty equivalent, which can help assess the relative importance of the cost of risk (R) compared to expected net revenue $[E(\pi)]$. Substituting the risk premium into the certainty equivalent (6) gives

$$CE_a = E(\pi) - R_a = E(\pi) - \tfrac{1}{2}\, r_2\, M_2 - 1/6\, r_3\, M_3. \qquad (7.7)$$

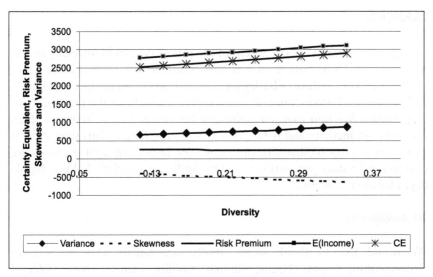

FIGURE 7-1 Certainty Equivalent, Mean Revenue, and Risk Premium. Simulations Results at Sample Means. (Values in Ethiopian *Birr.*)

This decomposes the certainty equivalent into three additive parts: (1) expected return $\{E(\pi)\}$, (2) variance ($R_{a2} \equiv \frac{1}{2} r_2 M_2$), and (3) skewness ($R_{a3} \equiv 1/6\ r_3 M_3$) components of the risk premium. Equation (7.7) provides a convenient way to investigate hypotheses concerning effects of crop diversity and land degradation on expected income, variance and skewness; these issues are explored with respect to barley in Tigray. To conserve space we do not present the simulation for cereals, though results are available from the authors.

Figure 7-1 presents our simulation results. We see that diversity increases mean revenues for the range of values, suggesting that it supports agricultural productivity. The risk premium R_a declines when diversity increases, indicating that diversity reduces the cost of uncertainty. The key question is whether either the second or third moments or *both* variance and skewness are responsible for this decline.

We find that diversity has a modest and positive effect on variance (R_{a2}), but this effect is dominated by the role of diversity in reducing the risk of crop failure; biodiversity therefore tends to increase variance, but it reduces downside risk when fertility is at the sample mean. The dominance of the skewness effect is shown by the certainty equivalent, which increases for all diversity values.

To gain additional insight, we now vary soil fertility. We find that when fertility is at the sample mean (medium fertility) the elasticity of the risk premium with respect to diversity is –0.124, but is –0.7 when fertility is low. This indicates that when the land is more fertile diversity has less effect on crop failure; biodiversity therefore particularly reduces risk on degraded lands.

Conclusions

This chapter presents an assessment of the role of crop of biodiversity in risk management. Using data from a survey conducted in Tigray, Ethiopia, we analyze the contribution of barley and cereal diversity on the mean, variance, and skewness of production; skewness captures effects on the probability of crop failure. We find that maintaining a larger number of barley varieties supports productivity and reduces the risk of crop failure. This suggests that maintaining crop biodiversity can be a very important strategy to secure harvests and reduce production risk. This finding particularly holds in highly degraded environments.

Acknowledgments

We would like to thank Randy Bluffstone, Anders Ekbom, Berhanu Gebremedhin, Fitsum Hagos, Gunnar Köhlin, Menale Kassie, Wilfred Nyangena, and Mahmud Yesuf for useful comments and suggestions. The usual disclaimer applies. Part of the material presented in this chapter relies upon paper by the same authors published in August 2009: "On Crop Biodiversity, Risk Exposure and Food Security in the Highlands of Ethiopia," *American Journal of Agricultural Economics* 91(3): 599–611.

Notes

1   Hurni (1993) estimated that 42 t/ha of soil were lost on the sloped cropland of Ethiopia each year.
2   It should be noted that aside from its "private" value, genetic diversity has an important social value, especially for plant breeding programs (Smale et al. 2001; Evenson and Gollin 1997).
3   Recently, the stochastic production function approach has been criticized by Chambers and Quiggin (2000), who suggested the adoption of the "state-contingent" approach to model production uncertainty.
4   Barrett et al. (2004) provide a farm level fixed effects by using plots information. This would not have been appropriate in the context of this chapter because we attempt to capture synergies at the farm level. Note, moreover, that farms in the sample do intercrop and grow one variety per plot. Therefore, a plot level analysis would not capture the benefit of crop genetic diversity.
5   The issue of "weak instruments" was also explored.
6   Typical estimates of relative aversion have varied between 1 and 5 (e.g., Binswanger 1981; Chavas and Holt 1996; Gollier 2001). Choosing a relative risk aversion parameter of 2.5 corresponds to a situation of moderate risk aversion.
7   Increasing downside risk means increasing the skewness of the distribution toward low outcomes, holding both mean and variance constant (Menezes et al. 1980).

References

Alemayehu, F. 1995. Genetic Variation between and within Ethiopian Barley Landraces with Emphasis on Durable Resistance. Ph.D. Thesis, Landbouw Universiteit, Wageningen, The Netherlands.

Antle, J. M. 1983. Testing the Stochastic Structure of Production: A Flexible Moment-Based Approach. *Journal of Business and Economic Statistics* 1: 192–201.

Asfaw, Z. 2000. The Barleys of Ethiopia. In *Genes in the Field*, edited by S. Brush. Boca Raton, FL: Lewis Publishers and Ottawa, Canada: International Development Research Centre.

Barrett, C. B., C. M. Moser, O. V. McHugh, and J. Barison. 2004. Better Technology, Better Plots, or Better Farmers? Identifying Changes in Productivity and Risk among Malagasy Rice Farmers. *American Journal of Agricultural Economics* 86: 869–88.

Binswanger, H.P. 1981. Attitudes toward Risk: Theoretical Implications of an Experiment in Rural India. *The Economic Journal* 91: 867–90.

Bromley, D. W., and J. P. Chavas. 1989. On Risk, Transactions, and Economic Development in the Semi-arid Tropics. *Economic Development and Cultural Change* 37: 719–36.

Chambers, R. G., and J. Quiggin. 2000. *Uncertainty, Production, Choice, and Agency: The State-Contingent Approach.* Cambridge: Cambridge University Press.

Chavas, J. P., and M. T. Holt. 1996. Economic Behavior under Uncertainty: A Joint Analysis of Risk Preferences and Technology. *Review of Economics and Statistics* 78: 329–35.

Davidson, R., and J. G. MacKinnon. 1993. *Estimation and Inference in Econometrics.* New York: Oxford University Press.

Deaton, A. 1989. Savings in Developing Countries: Theory and Review. World Bank Economic Review. Proceedings of the Annual World Bank Conference on Development Economics 61–96.

———. 1992. Saving and Income Smoothing in Côte d'Ivoire. *Journal of African Economies* 1: 1–24.

Dercon, S. 1996. Risk, Crop Choice, and Savings: Evidence from Tanzania. *Economic Development and Cultural Change* 44(3): 385–415.

———. 2004. Growth and Shocks: Evidence from Rural Ethiopia. *Journal of Development Economics* 74(2): 309–29.

———. 2005. Insurance Against Poverty. World Institute for Development Economics Research, United Nations University, Helsinki.

Dercon, S., and P. Krishnan. 2000. Vulnerability, Poverty, and Seasonality in Ethiopia. *Journal of Development Studies* 36(6): 25–53.

Di Falco, S., and C. Perrings. 2005. Crop Biodiversity, Risk Management, and the Implications of Agricultural Assistance. *Ecological Economics* 55: 459–66.

Di Falco, S., and J.-P. Chavas. 2009. On Crop Biodiversity, Risk Exposure, and Food Security in the Highlands of Ethiopia. *American Journal of Agricultural Economics* 91(3): 599–611.

Evenson, R. E., and D. Gollin. 1997. Genetic Resources, International Organizations, and Improvement in Rice Varieties. *Economic Development and Cultural Change* 45: 471–500.

Fafchamps, M. 1992. Cash Crop Production, Food Price Volatility, and Rural Market Integration in the Third World. *American Journal of Agricultural Economics* 74: 90–99.

Fafchamps, M., and J. Pender. 1997. Precautionary Savings, Credit Constraints, and Irreversible Investment: Evidence from Semi-arid India. *Journal of Business and Economic Statistics* 15: 180–94.

Fafchamps, M., C. Udry, and K. Czukas. 1998. Drought and Saving in West Africa: Are Livestock a Buffer Stock? *Journal of Development Economics* 55: 273–305.

FAO (Food and Agriculture Organization of the United Nations). 2001. *The Economics of Soil Productivity in Sub-Saharan Africa.* Rome.

Gollier, C. 2001. *The Economics of Risk and Time.* Cambridge, MA: MIT Press.

Grepperud, S. 1996. Population Pressure and Land Degradation: The Case of Ethiopia. *Journal of Environment, Economics, and Management* 30: 18–33.

Harlan, J. R. 1992. *Crops and Man,* 2nd ed. Madison, Wisconsin: American Society of Agronomy and Crop Science Society of America.

Heal, G. 2000. *Nature and the Marketplace: Capturing the Value of Ecosystem Services.* New York: Island Press.

Hurni, H. 1993. Land Degradation, Famines, and Resource Scenarios in Ethiopia. In *World Soil Erosion and Conservation,* edited by D. Pimental. Cambridge: Cambridge University Press, 27–62.

Just, R. E., and R. D. Pope. 1979. Production Function Estimation and Related Risk Considerations. *American Journal of Agricultural Economics* 61: 276–84.

Just, R. E., and W. Candler. 1985. Production Functions and Rationality of Mixed Cropping. *European Review of Agricultural Economics* 12: 207–31.

Kurosaki, T., and M. Fafchamps. 2002. Insurance Market Efficiency and Crop Choices in Pakistan. *Journal of Development Economics* 67: 419–53.

Menezes, C., C. Geiss, and J. Tressler. 1980. Increasing Downside Risk. *American Economic Review* 70: 921–32.

Negassa, M. 1985. Geographic Distribution and Genotypic Diversity of Resistance to Powdery Mildew of Barley of Ethiopia. *Hereditas* 102: 139–50.

Paxson, C. H. 1992. Using Weather Variability to Estimate the Response of Saving to Transitory Income in Thailand. *American Economic Review* 82: 15–33.

Pender, J., and B. Gebremedhin. 2004. Impacts of Policies and Technologies in Dryland Agriculture: Evidence from Northern Ethiopia. In *Challenges and Strategies for Dryland Agriculture.* CSSA Special Publication No. 32, Crop Science Society of America and American Society of Agronomy, Madison, Wisconsin, 389–416.

Pender, J., F. Place, and S. Ehui. 1999. Strategies for Sustainable Agricultural Development in the East African Highlands. Environment and Production Technology Division Working Paper 41. Washington, DC: International Food Policy Research Institute.

Pratt, J. W. 1964. Risk Aversion in the Small and in the Large. *Econometrica* 32: 122–36.

Qualset C. O. 1975. Sampling Germplasm in a Centre of Diversity: An Example of Disease Resistance in Ethiopian Barley. In *Crop Genetic Resources for Today and Tomorrow,* edited by O. H. Frankel and G. J. Hawks. Cambridge: Cambridge University Press.

Qualset, C. O., and J. G. Moseman. 1966. Disease Reaction of 654 Barley Introduction from Ethiopia. USDA/ARS Progress Report, Unpublished.

REST (Relief Society of Tigray) and NORAGRIC at the Agricultural University of Norway. 1995. Farming Systems, Resource Management, and Coping Strategies in Northern Ethiopia: Report of a Social and Agro-ecological Baseline Study in Central Tigray.

Smale, M., J. Hartell, P. W. Heisey, and B. Senauer. 1998. The Contribution of Genetic Resources and Diversity to Wheat Production in the Punjab of Pakistan. *American Journal of Agricultural Economics* 80: 482–93.

Smale, M, M. R. Bellon, A. Gómez, and J. Alfonso. 2001. Maize Diversity, Variety Attributes, and Farmers' Choices in Southeastern Guanajuato, Mexico. *Economic Development and Cultural Change* 50: 201–25.

Tesemma, T., and E. Bechere. 1998. Developing Elite Durum Wheat Landrace Selections (Composites) for Ethiopian Peasant Farm Use: Raising Productivity While Keeping Diversity Alive. *Euphytica* 102: 323–28.

Vavilov, N. I. 1949. The Origin, Variation, Immunity, and Breeding of Cultivated Plants. *Chronica Botanica* 13: 1–364.

Widawsky, D., and S. Rozelle. 1998. Varietal Diversity and Yield Variability in Chinese Rice Production. In *Farmers, Gene Banks, and Crop Breeding*, edited by M. Smale. Boston: Kluwer.

Wooldridge, J. 2002. *Econometric Analysis of Cross Section and Panel Data*. Cambridge, MA: MIT Press.

PART III

Returns to Sustainable Land Management Investments

Returns to Sustainable Land Management Investments

CHAPTER 8

Where Does Investment on Sustainable Land Management Technology Work? Empirical Evidence from the Ethiopian Highlands

MENALE KASSIE

M any of the first seven chapters in this volume discuss the importance and consequences of land degradation for food production in East Africa. As presented in Part 1 on determinants of sustainable land management (SLM) investments, development organizations have promoted SLM technologies such as soil and water conservation (SWC) and reduced tillage. Unfortunately, adoption and success of such technologies has been limited (Amsalu and de Graaff 2007; Shiferaw and Holden 2001), with a common explanation being that SLM investments may not be profitable for average farmers.[1] Analyzing returns of too often very costly SLM investments is therefore of primary importance.

The literature suggests that economic performance of SLM technologies depends on a number of factors, including agro-ecology, market development, availability of supporting institutions, and farm-specific characteristics. Several researchers have analyzed the adoption and impact of SWC practices on productivity (Byiringiro and Reardon 1996; Shively 1999, 1998a, 1998b; Holden et al. 2001; Bekele 2005; Benin 2006; Kassie and Holden 2006; Pender and Gebremedhin 2007), but studies of the relationship between reduced tillage and crop productivity are very limited as are analyses of agro-ecological effects. To the best of our knowledge only Pender and Gebremedhin (2007) investigate the impacts of reduced tillage on productivity.

The objective of this chapter is to use plot-level data to examine the economic performance of stone bunds and reduced tillage in low rainfall areas in Tigray Regional State and higher rainfall areas of Amhara Regional State in the highlands of northern Ethiopia. The next section of the chapter reviews the empirical approach. We then present the data, followed by results. The final section of the chapter concludes and attempts to draw out the key policy implications.

Empirical Approach

This chapter seeks to extend the existing literature by addressing some important methodological limitations. A critical issue with previous studies (e.g.,
Byiringiro and Reardon, 1996; Shively, 1999; 1998a; 1998b; Holden et al.,
2001; Bekele 2005; Kaliba and Rabele 2004; Benin 2006; Kassie and Holden
2006; Pender and Gebremedhin 2007) evaluating the impact of soil conservation technologies at the farm level is that in many cases they do not properly
control for potential differences between technology adopters (participants)
and farmers in comparison groups (non-participants), making it difficult to
draw definitive conclusions. Farmers who adopt conservation technologies
are likely to be different from other farmers in ways that are unobservable.

For example, if more motivated farmers are more likely to be selected,
comparing adopters to other farmers would overestimate a technology's
impact on farm-level outcomes. Second, individuals might be chosen by
development agencies based on their propensities to adopt, but non-random
selection of farm households likely implies simple comparisons of outcomes
between adopters and non-adopters and is likely to yield biased estimates of
technology impacts.

In addition, most previous studies use pooled ordinary least squares
(OLS) regression, regressing farm-level outcomes (in our case, value of crop
production per hectare) on adoption and other variables. Pooled estimation
assumes that covariates have the same impact on adopters as non-adopters
(i.e., common slope coefficients). This implies that SLM adoption has only
an intercept shift effect that is the same regardless of the values taken by other
covariates. These strong assumptions are commonly used in the literature (e.g.,
Byiringiro and Reardon 1996; Shively 1998a, 1998b; Pender and Gebremedhin
2007) without testing, but in this chapter such assumptions are not presumed
to hold. Indeed, in our sample a Chow test of equality of coefficients for adopters and non-adopters of stone bunds and reduced tillage rejected equality at
the 1.0% significance level. We therefore use a regression approach that differ-
entiates coefficients for adopters and non-adopters.

In this chapter I use propensity score matching (PSM) and switching
regression to address econometric issues and ensure robustness. The rationale
behind PSM is that one group participates in a program or treatment (e.g.,
adopting a technology), while another does not. The objective is then to assess
the effectiveness of the treatment by comparing average outcomes, allowing
for the possibility that adoption of SLM could be non-random, which is especially relevant when observational rather than experimental data are used.

We match plots and associated households that do and do not have SLM
technologies based on characteristics that are similar and relevant to technology choice. This approach reduces the potential for bias, although there may
still be selection bias caused by differences in unobservables. PSM also allows
us to analyze observations that fall within common support using switching

regression. The matching process has two steps. First, we use a probit model to estimate the propensity scores and then use nearest-neighbor (NN) matching based on propensity score estimates to calculate the average treatment on the treated (ATT). Compared to other weighted matching methods, such as kernel matching, NN allows us to identify the specific matched observations that enter the calculation of the ATT, which we then use for parametric regressions.

Study Sites and Data

In this chapter I use data from community, household and plot level surveys conducted in areas at or above 1,500 meters above sea level in the Amhara and Tigray regions between 1999 and 2000. The Amhara data set includes 435 households, 98 villages, 49 peasant associations, and about 1,365 plots, while the Tigray data include 500 households, 100 villages, 50 peasant associations, and 1,032 plots. Information was collected about household structure and endowments, access to infrastructure and services, plot characteristics (e.g., size, slope, quality, mode of acquisition, soil texture and color, altitude), land investment, land management practices, inputs, and agricultural production. Primary data are supplemented by secondary information on rainfall and population.

Tables 8-1 and 8-2 present descriptive statistics by region for the subsamples of plots with and without stone bunds and reduced tillage. About 37% of sample plots in Tigray and 17% percent in Amhara have stone bunds. Reduced tillage is used less often, with 13% of plots in Tigray and 15% in Amhara employing that method. Average annual rainfall in Amhara is about 1,980 mm per year and 650 mm in Tigray. Rainfall in Amhara therefore averages about three times that of Tigray.[2] At 142 to 144 persons per square kilometer, mean population density is similar across the two regional samples.

Between 1991 and 2000 about 64% of stone bunds in Amhara and 37% in Tigray (1997 to 1999) were constructed using private resources. Government extension workers and peasant association officials mobilized community labor for construction of 29% of bunds in Amhara and 55% of bunds in Tigray. In Tigray a combination of private and mass mobilization investment was used for 4% of stone bunds and the remainder with other support, such as food-for-work. There is no statistically significant mean yield difference between plots built with private and mass mobilization resources, though average yield is higher on plots with private investments.

We find that the mean value of crop production per hectare is $300 on Tigray plots with reduced tillage compared with $220 on standard plots. Comparable values in Amhara are $241 and $222, suggesting that reduced tillage may increase yields. For stone bunds in Tigray we find mean yields of $239 on conserved plots compared with $201 on standard plots. The picture is more mixed in Amhara, however, where plots with bunds yield an average $226 versus $230 on non-conserved plots. We emphasize, however, that this

TABLE 8-1 Descriptive Statistics of Variables for the Tigray Region

Variables	With Reduced Tillage	Without Reduced Tillage	With Stone Bunds	Without Stone Bunds
Gross crop revenue [gross crop value production], in ETB/hectare (ETB = Ethiopian *birr*)	2099.625 (2188.190)	1542.611 (1708.732)	1670.375 (1713.941)	1409.338 (1509.163)
Market distance [distance of residence to markets], in walking hours	3.317 (2.925)	3.247 (2.610)	3.129 (2.399)	2.983 (2.495)
Gender [sex of household head], 1 = male; 0 = female)	0.840	0.830	0.913	0.886
Age [age of household head], in years	47.580 (12.466)	46.745 (14.623)	49.684 (12.173)	49.673 (12.165)
Family size [number of household members]	5.634 (2.084)	5.702 (2.194)	6.140 (1.991)	6.188 (2.193)
Illiterate [household head is illiterate], 1 = yes; 0 = otherwise	0.893	0.851	0.862	0.842
Education low [household head has schooling to grades 1 and 2], 1 = yes; 0 = otherwise	0.069	0.096	0.092	0.079
Education high [household head has schooling above grade 3], 1 = yes; 0 = otherwise	0.038	0.053	0.046	0.079
Oxen [number of oxen owned by household]	1.267 (0.910)	1.415 (0.944)	1.327 (0.873)	1.401 (0.818)
Other cattle [number of cattle other than oxen]	3.992 (4.244)	3.809 (3.514)	3.194 (3.432)	3.54 (3.353)
Small ruminants [number of small ruminant animals]	9.450 (11.916)	7.713 (12.414)	5.724 (8.529)	6.134 (9.064)
Pack animal [number of pack animals]	1.130 (1.786)	1.000 (1.698)	0.921 (1.511)	1.114 (1.432)
Extension contact [household has extension contact], 1 = yes; 0 = otherwise	0.198	0.245	0.196	0.178
Farm size [total land holdings], in hectares	1.785 (2.276)	1.284 (0.718)	1.097 (1.038)	1.182 (0.654)
Plot area, in hectares	0.322 (0.261)	0.319 (0.251)	0.304 (0.224)	0.304 (0.253)
Fertilizer dummy [fertilizer use on plot], 1 =yes; 0 = otherwise	0.176	0.234	0.329	0.297
High fertile soil [high fertile soil plot], 1= yes; 0= otherwise	0.099	0.064	0.105	0.129
Medium fertile soil [moderate fertile soil plot], 1= yes; 0 = otherwise	0.618	0.649	0.625	0.609

TABLE 8-1 *(Cont.)*

Variables	With Reduced Tillage	Without Reduced Tillage	With Stone Bunds	Without Stone Bunds
Infertile soil [infertile soil plot], 1 = yes; 0= otherwise	0.282	0.287	0.270	0.262
Deep soil [deep plot soil depth], 1 = yes; 0 = otherwise	0.168	0.170	0.145	0.173
Moderately deep soil [moderately deep plot soil depth], 1 = yes; 0 = otherwise	0.344	0.298	0.357	0.381
Shallow soil [shallow plot soil depth], 1 = yes; 0 = otherwise	0.489	0.532	0.497	0.446
Black soil [black soil in plot], 1 = yes; 0 = otherwise	0.099	0.128	0.171	0.218
Brown soil [brown soil in plot], 1 = yes; 0 = otherwise	0.229	0.213	0.191	0.134
Gray soil [gray soil in plot], 1 = yes; 0 = otherwise	0.321	0.309	0.253	0.277
Red soil [red soil in plot], 1 = yes; 0 = otherwise	0.351	0.351	0.385	0.371
Flat slope [flat plot slope], 1 = yes; 0 = otherwise	0.603	0.638	0.439	0.579
Moderate slope [moderate plot slope], 1 = yes; 0 = otherwise	0.313	0.234	0.416	0.351
Steep slope [steep plot slope], 1= yes; 0 = otherwise	0.084	0.128	0.145	0.069
Clay soil [clay soil in plot], 1 = yes; 0 = otherwise	0.130	0.160	0.161	0.203
Loam soil [loam soil in plot], 1 = yes; 0 = otherwise	0.565	0.447	0.418	0.347
Sandy soil [sandy soil in plot], 1 = yes; 0 = otherwise	0.229	0.309	0.304	0.332
Silt soil [silt soil in plot], 1 = yes; 0 = otherwise	0.076	0.085	0.117	0.119
No erosion [plot not eroded], 1 = yes; 0 = otherwise	0.649	0.660	0.548	0.653
Moderate erosion [plot moderately eroded], 1 = yes; 0 = otherwise	0.267	0.309	0.355	0.297
Severe erosion [plot severely eroded], 1 = yes; 0 = otherwise	0.084	0.032	0.097	0.050
Gully plot [Gully on plot], 1 = yes; 0 = otherwise	0.038	0.021	0.041	0.030

(continued)

TABLE 8-1 *(Cont.)*

Variables	With Reduced Tillage	Without Reduced Tillage	With Stone Bunds	Without Stone Bunds
Plot distance [distance from residence to plot], in hours walking	0.352 (0.487)	0.305 (0.348)	0.281 (0.375)	0.307 (0.341)
Rented plot, 1 = yes; 0 = otherwise	0.107	0.117	0.097	0.089
Irrigation [plot irrigated], 1 = yes; 0 = otherwise	0.053	0.011	0.008	0.010
Altitude [village altitude], in meters above sea level	2120.771 (436.477)	2113.745 (328.509)	2171.408 (316.787)	2123.099 (319.476)
Rainfall [mean rainfall], in mm	635.309 (95.806)	659.573 (95.131)	653.244 (93.020)	649.394 (92.692)
Population density [village population density], in person/km²	131.213 (72.430)	132.147 (65.716)	151.528 (75.016)	142.403 (67.491)
Southern zone	0.237	0.213	0.242	0.262
Central zone	0.427	0.468	0.526	0.436
Eastern zone	0.275	0.202	0.143	0.198
Western zone	0.061	0.117	0.089	0.104
Soil and water conservation [SWC structures on plot], 1 = yes; 0 = otherwise	0.504	0.468		
Reduced tillage [Reduced tillage on plot], 1 = yes; 0 = otherwise			0.130	0.129
Number of observations	131	94	392	202

TABLE 8-2 Descriptive Statistics of Variables for the Amhara Region

Variables	With Stone Bunds	Without Stone Bunds	With reduced Tillage	Without Reduced Tillage
Gross crop revenue [gross crop value production], in ETB/hectare (ETB = Ethiopian *birr*)	1582.315 (1115.503)	1612.601 (1706.398)	1688.400 (1646.089)	1557.364 (1473.371)
Slope [degree of plot slope]	8.030 (5.590)	6.915 (7.285)	7.613 (7.897)	7.503 (6.637)
Red soil [red soil in plot], 1 = yes; 0 = otherwise	0.197	0.197	0.226	0.235
Black soil [black soil in plot], 1 = yes; 0 = otherwise	0.363	0.380	0.417	0.430
Brown soil [black soil in plot], 1 = yes; 0 = otherwise	0.359	0.338	0.276	0.268
Gray soil [gray soil in plot], 1 = yes; 0 = otherwise	0.085	0.085	0.080	0.067

TABLE 8-2 *(Cont.)*

Variables	With Stone Bunds	Without Stone Bunds	With reduced Tillage	Without Reduced Tillage
Deep soil [deep plot soil depth], 1 = yes; 0 = otherwise	0.137	0.162	0.090	0.128
Moderately deep soil [moderate plot soil depth], 1 = yes; 0 = otherwise	0.594	0.613	0.578	0.617
Shallow soil [shallow plot soil depth], 1 = yes; 0 = otherwise	0.256	0.225	0.312	0.242
No erosion [plot not eroded], 1 = yes; 0 = otherwise	0.675	0.620	0.588	0.550
Moderate erosion [plot moderately eroded], 1 = yes; 0 = otherwise	0.556	0.507	0.462	0.416
Severe erosion [plot severely eroded], 1 = yes; 0 = otherwise	0.111	0.113	0.116	0.134
Silt soil [silt soil in plot], 1 = yes; 0 = otherwise	0.410	0.338	0.377	0.362
Clay soil [clay soil in plot], 1 = yes; 0 = otherwise	0.077	0.056	0.095	0.094
Loam soil [loam soil in plot], 1 = yes; 0 = otherwise	0.363	0.423	0.357	0.383
Sandy soil [sandy soil in plot], 1 = yes; 0 = otherwise	0.145	0.183	0.161	0.161
Infertile soil [infertile soil in plot], 1 = yes; 0 = otherwise	0.218	0.211	0.256	0.221
Highly fertile soil [highly fertile soil in plot], 1 = yes; 0 = otherwise	0.038	0.035	0.045	0.054
Medium fertile soil [moderately fertile soil in plot], 1 = yes; 0 = otherwise	0.726	0.739	0.688	0.718
Plot distance [distance from residence to plot[, in hours walking	0.254 (0.275)	0.268 (0.346)	0.458 (1.803)	0.362 (0.753)
Rented plot, 1 = yes; 0 = otherwise	0.051	0.063	0.065	0.074
Irrigation [plot irrigated], 1 = yes; 0 = otherwise	0.021	0.000	0.035	0.047
Manure [plot manured]], 1 = yes; 0 = otherwise	0.090	0.085	0.070	0.081
Fertilizer dummy [fertilizer use on plot], 1 = yes; 0 = otherwise	0.184	0.148	0.216	0.181
Plot area, in hectares	0.431 (0.275)	0.463 (0.386)	0.416 (0.328)	0.478 (0.425)

(continued)

TABLE 8-2 *(Cont.)*

Variables	With Stone Bunds	Without Stone Bunds	With reduced Tillage	Without Reduced Tillage
Market distance [distance of residence to markets], in walking hours	1.432 (1.133)	1.410 (1.176)	1.647 (1.288)	1.624 (1.270)
Gender [sex of household head], 1 = male; 0 = female	0.983	0.958	0.945	0.940
Family size [number of household members]	6.573 (2.156)	6.803 (2.325)	6.809 (2.087)	6.718 (2.474)
Age [age of household head], in years	47.120 (12.146)	46.437 (12.571)	44.186 (10.419)	44.181 (11.091)
Education [household head education level]	2.641 (3.220)	2.331 (3.096)	2.839 (3.299)	2.577 (3.025)
Extension contact	0.705	0.634	0.613	0.570
Population density [village population], in person/km²	141.957 (87.935)	137.169 (78.228)	118.779 (75.109)	124.685 (62.116)
Rainfall [mean rainfall], in mm	1843.835 (603.785)	1911.819 (683.966)	(711.869)	2003.145 (671.875)
North Gonder zone	0.094	0.120	0.236	0.208
South Gonder zone	0.137	0.148	0.111	0.128
East Gojam zone	0.047	0.092	0.015	0.034
West Gojam zone	0.004	0.014	0.040	0.013
Awi zone	0.000	0.000	0.005	0.013
North Shewa zone	0.175	0.183	0.156	0.168
North Wello	0.124	0.113	0.121	0.101
South Wello zone	0.419	0.331	0.317	0.336
Livestock [number of livestock], in tropical livestock units	2.331 (1.778)	2.653 (2.701)	2.403 (1.969)	2.120 (2.038)
Farm size [total land holdings], in hectares	1.392 (0.715)	1.499 (1.036)	1.525	1.560 (1.152)
Soil and water conservation [SWC structures on plot], 1 = yes; 0 = otherwise			0.362	0.322
Reduced tillage [Reduced tillage on plot], 1 =yes; 0= otherwise	0.286	0.275		
Number of observations	234	142	199	149

output difference may not be the result of soil bunds and reduced tillage, but instead may be due to factors such as cropland quality, input use or other features. Careful multivariate analysis is therefore called for.

Results and Discussion

In this section, we present our results using the PSM and OLS switching regression methods.

Propensity Score Matching Method

Tables 8-3 and 8-4 present probit results that explore how plot and household characteristics influence decisions to adopt stone bunds and reduced tillage. The sample passes standard balancing tests, such as those described in Becker and Ichino (2002). In the interest of space, and because our main goal

TABLE 8-3 Adoption of Reduced Tillage and Stone Bund in Tigray Region

Explanatory Variables	Reduced Tillage	Stone Bunds
Medium fertile plots	−0.025	−0.083
	(0.235)	(0.176)
Infertile soil plots	−0.137	−0.369*
	(0.279)	(0.214)
Medium soil	0.127	0.198
	(0.201)	(0.153)
Shallow soil	0.439**	0.480***
	(0.219)	(0.173)
Moderate slope	−0.268*	0.604***
	(0.142)	(0.108)
Steep slope	−0.466*	0.670***
	(0.243)	(0.180)
Brown soil	0.534**	0.217
	(0.258)	(0.195)
Gray soil	0.409	0.077
	(0.266)	(0.190)
Red soil	0.412	−0.171
	(0.254)	(0.181)
Loam soil	0.153	0.108
	(0.237)	(0.182)
Clay soil	−0.096	0.179
	(0.250)	(0.184)
Sandy soil	−0.065	0.424*
	(0.298)	(0.222)
Moderate erosion	−0.016	0.208*
	(0.141)	(0.109)
Severe erosion	−0.149	0.238
	(0.253)	(0.197)

(continued)

TABLE 8-3 *(Cont.)*

Explanatory Variables	Reduced Tillage	Stone Bunds
Plot distance from residence	0.083	−0.351**
	(0.171)	(0.146)
Rented plot	0.009	−0.184
	(0.179)	(0.144)
Gully plot	0.140	0.092
	(0.306)	(0.256)
Farm size	0.683***	−0.039
	(0.129)	(0.056)
Manure plot	−0.101	0.327***
	(0.148)	(0.112)
Irrigation	0.555*	−0.862**
	(0.307)	(0.368)
Stone-covered plot	0.343**	0.393***
	(0.143)	(0.115)
Ln(population density)	0.028	0.286**
	(0.148)	(0.115)
Ln(rainfall)	0.899	−0.550
	(0.701)	(0.524)
Market distance	−0.038	0.029
	(0.029)	(0.023)
Gender	−0.367*	0.170
	(0.199)	(0.169)
Ln(age)	−0.187	0.216
	(0.201)	(0.165)
Family size	−0.118***	0.045*
	(0.033)	(0.024)
Education low	−0.047	0.312*
	(0.231)	(0.181)
Education high	−0.465	−0.164
	(0.289)	(0.209)
Extension contact	0.372**	0.047
	(0.157)	(0.129)
Oxen	−0.173**	−0.101*
	(0.072)	(0.057)
Other cattle	0.025	−0.036**
	(0.020)	(0.017)
Small ruminants	0.016***	−0.004
	(0.005)	(0.005)
Pack animals	0.007	0.033
	(0.046)	(0.040)
Ln(Plot area)	−0.093	0.187***
	(0.085)	(0.063)

TABLE 8-3 *(Cont.)*

Explanatory Variables	Reduced Tillage	Stone Bunds
Fertilizer dummy	−0.490***	
	(0.142)	
Soil and water conservation	0.086	
	(0.128)	
Reduced tillage		0.043
		(0.146)
Joint significance of zonal variables	6.56*	20.99***
_cons	−6.984	0.647
	(4.762)	(3.533)
Pseudo R2	0.2084	0.2036
Model chi-square	163.711***	272.130***
Number of observations	1032	997

Note: ***, **, * is significant at 1%, 5%, and 10%, respectively

TABLE 8-4 Adoption of Reduced Tillage and Stone Bund in Amhara Region

Explanatory Variables	Reduced Tillage	Stone Bunds
Slope	0.013	0.034***
	(0.009)	(0.008)
Black soil	0.242*	0.018
	(0.143)	(0.145)
Brown soil	−0.071	−0.016
	(0.147)	(0.146)
Gray soil	−0.050	−0.084
	(0.212)	(0.202)
Deep soil	−0.600***	−0.145
	(0.184)	(0.174)
Moderately deep soil	−0.115	0.092
	(0.131)	(0.133)
Moderate erosion	0.179	0.547***
	(0.117)	(0.111)
Severe erosion	−0.244	0.126
	(0.192)	(0.176)
Clay soil	−0.159	−0.327*
	(0.182)	(0.188)
Loam soil	−0.163	−0.165
	(0.118)	(0.115)
Sandy soil	0.039	−0.173
	(0.160)	(0.158)

(continued)

TABLE 8-4 *(Cont.)*

Explanatory Variables	Reduced Tillage	Stone Bunds
High fertile soil	−0.012 (0.248)	−0.123 (0.261)
Medium fertile soil	0.035 (0.136)	0.155 (0.135)
Plot distance	0.036 (0.053)	−0.285** (0.142)
Rented plot	0.070 (0.199)	−0.001 (0.207)
Irrigation	0.319 (0.255)	−0.528* (0.307)
Manure plot	−0.026 (0.185)	−0.093 (0.182)
Ln(plot are)	−0.012 (0.082)	0.414*** (0.085)
Market distance	0.027 (0.049)	−0.047 (0.052)
Gender	−0.490** (0.226)	0.670** (0.293)
Family size	0.045* (0.026)	0.008 (0.026)
Age	−0.012** (0.005)	0.007 (0.005)
Education	0.006 (0.017)	−0.000 (0.016)
Extension contact	−0.161 (0.106)	0.299*** (0.109)
Population density	−0.004*** (0.001)	−0.322*** (0.107)
Ln(rainfall)	−0.064 (0.191)	−0.521*** (0.179)
Livestock	−0.035 (0.028)	−0.016 (0.027)
Farm size	−0.006 (0.059)	−0.256*** (0.074)
Soil and water conservation	0.083 (0.123)	
Fertilizer dummy	0.361** (0.142)	
Reduced tillage		0.098 (0.128)
Joint significant test of zonal dummy variables	66.39***	59.17***

TABLE 8-4 *(Cont.)*

Explanatory Variables	Reduced Tillage	Stone Bunds
Constant	1.391	3.798**
	(1.559)	(1.605)
LR chi2	298.446***	355.540***
Pseudo R2	0.2632	0.2907
Number of observations	1365	1294

Note: ***, **, * is significant at 1%, 5%, and 10%, respectively.

is to identify the average treatment effect and use matched observations in switching regressions, estimates are not discussed in detail. Results in Tables 8-3 and 8-4 show that both socioeconomic and plot characteristics are significant for adoption.

Table 8-5 provides the NN matching estimates of crop production per hectare. The results reveal that reduced tillage increases crop productivity by $86 and stone bunds improve yields by $53 per hectare in Tigray. Both conservation methods are therefore found to improve yields in low-rainfall Tigray.

In Amhara, which has three times higher average rainfall than Tigray, neither reduced tillage nor stone bunds has a statistically significant impact on crop productivity. Indeed, the estimated yield increment when stone bunds are used is actually negative (−$22.00 per hectare), suggesting that stone bunds may reduce yields in higher rainfall environments. This result is likely due to water logging and other negative technological effects of stone bunds that are more likely to occur in higher rainfall regions.[3]

TABLE 8-5 Productivity Impacts from Semi-parametric Regression Analysis

	Reduced Tillage vs. Non-reduced Tillage Plots	Stone Bund Plots vs. Non-stone Bund Plots
ATT	19.794	−156.324
Standard error	189.744	272.382
Treated	199	234
Control	149	142
ATT	604.839**	369.807***
Standard error	317.296	142.510
Treated	131	392
Control	94	202

Note: ** significant at 5%; *** significant at 1%
* Bootstrapped standard errors were used to take into account the estimated propensity score used in the second stage (nearest-neighbor matching estimator). ATT = average treatment effect on the treated.

Switching Regression Estimates

Switching regression is used to further investigate any crop production gaps between plots with stone bunds, reduced tillage, and neither technology. Because we have data on multiple plots for each household we apply random effects on matched observations that are similar in the distribution of propensity scores and covariates. In the interest of brevity we do not discuss the regression results in detail; they are displayed in Tables 8-6 and 8-7.

Consistent with results from the PSM analysis, we find that stone bunds and reduced tillage lead to significantly higher productivity gains in low-rainfall Tigray compared with higher-rainfall Amhara (Table 8-8). Results from the two methods correspond closely, so we therefore conclude that stone bunds and reduced tillage are more productive in low-rainfall areas. We believe this is due to benefits of moisture conservation, while in high-rainfall areas the technologies may contribute to water logging, weeds, and pests. Conservation measures like moisture drainage ditches may protect soils during extreme rainfall events and also could be more appropriate.

Conclusions and Implications

As discussed throughout this volume, while land degradation continues to present serious challenges to crop production in Ethiopia, efforts to promote the adoption of SLM technologies have often been frustrated by low adoption and lack of sustainability. This chapter uses detailed plot-level data to examine the productivity gains associated with adoption of stone bunds and reduced tillage in the Ethiopian highlands. Our findings suggest that a clear understanding of the role of agro-ecology, such as rainfall, is important when promoting SLM.

We estimate the average treatment effect by utilizing semi- and non-parametric matching and parametric regression approaches. We use these different methods to try to ensure results are robust to specification and methodology. The parametric regression estimates are based on matched samples obtained from nearest-neighbor propensity score matching. This approach is important, because conventional regression estimates are obtained without ensuring that there actually exist comparable conserved and non-conserved plots on the distribution of covariates.

The estimates from the two methods tell a consistent story. Stone bunds have a positive significant yield impact in low rainfall areas. However, this impact is not observed in higher rainfall areas. Productivity impacts of SLM technologies are therefore likely to be agro-ecology specific. This finding highlights the importance of developing and disseminating agro-ecology specific soil conservation technologies that actually increase agricultural productivity. For instance, in high rainfall areas moisture conservation using physical

TABLE 8-6 Reduced Tillage and Stone Bunds Productivity Analysis Using Switching Regression: Tigray Region

Explanatory variables	With Reduced Tillage	Without Reduced Tillage	With Stone Bunds	Without Stone Bunds
Medium fertile plot	0.004	−0.069	−0.130	−0.010
	(0.274)	(0.292)	(0.149)	(0.181)
Infertile soil plot	0.068	−0.424	−0.320*	−0.049
	(0.336)	(0.370)	(0.177)	(0.224)
Deep soil	0.128	−0.726**	−0.036	−0.282
	(0.277)	(0.323)	(0.129)	(0.180)
Medium soil	−0.039	−0.417	−0.164	−0.391*
	(0.254)	(0.329)	(0.148)	(0.208)
Moderate slope	−0.071	−0.169	−0.195**	−0.079
	(0.206)	(0.161)	(0.083)	(0.117)
Steep slope	−0.124	0.309	−0.156	−0.358
	(0.411)	(0.274)	(0.110)	(0.228)
Brown soil	0.095	1.368***	0.037	0.205
	(0.302)	(0.451)	(0.149)	(0.253)
Gray soil	0.303	0.603	−0.031	0.339
	(0.359)	(0.390)	(0.163)	(0.271)
Red soil	0.237	1.038***	0.079	0.238
	(0.321)	(0.387)	(0.145)	(0.258)
Loam soil	−0.116	−0.801**	−0.054	0.025
	(0.316)	(0.374)	(0.132)	(0.320)
Clay soil	−0.569*	−0.946**	0.058	−0.346
	(0.325)	(0.444)	(0.149)	(0.284)
Sandy soil	−0.421	−1.062**	−0.098	−0.125
	(0.427)	(0.499)	(0.164)	(0.301)
Moderate erosion	−0.153	−0.099	0.004	0.158
	(0.188)	(0.195)	(0.083)	(0.143)
Severe erosion	−0.196	−1.68***	0.042	0.002
	(0.215)	(0.625)	(0.140)	(0.251)
Plot distance	−0.179	0.178	−0.182	−0.058
	(0.188)	(0.242)	(0.130)	(0.154)
Rented plot	−0.279	0.326	−0.122	0.268
	(0.230)	(0.258)	(0.108)	(0.170)
Gully plot	0.116	0.071	0.110	−0.406
	(0.303)	(0.397)	(0.227)	(0.281)
Farm size	0.014	0.235	−0.003	0.071
	(0.027)	(0.166)	(0.030)	(0.118)
Manure plot	0.079	0.002	0.253***	0.137
	(0.202)	(0.210)	(0.090)	(0.131)
Irrigation	1.119**	−0.316	0.028	0.086
	(0.488)	(0.409)	(0.235)	(0.294)

(continued)

Menale Kassie

TABLE 8-6 *(Cont.)*

Explanatory variables	With Reduced Tillage	Without Reduced Tillage	With Stone Bunds	Without Stone Bunds
Stone-covered plot	0.281*	−0.009	−0.010	0.110
	(0.153)	(0.177)	(0.097)	(0.104)
Ln(pop. density)	−0.163	0.133	−0.069	−0.004
	(0.199)	(0.203)	(0.124)	(0.166)
Ln(rainfall)	1.988*	1.210	0.676	−0.283
	(1.205)	(0.847)	(0.507)	(0.584)
Market distance	−0.124***	−0.119***	−0.054**	−0.067**
	(0.039)	(0.042)	(0.023)	(0.029)
Gender	−0.108	0.843***	0.258	0.477**
	(0.262)	(0.291)	(0.229)	(0.200)
Ln(Age)	−0.078	−0.209	−0.052	−0.288
	(0.321)	(0.246)	(0.187)	(0.205)
Family size	−0.023	−0.502**	−0.017	−0.025
	(0.202)	(0.220)	(0.030)	(0.036)
Education low	0.138	0.445	−0.024	0.171
	(0.337)	(0.305)	(0.169)	(0.217)
Education high	−0.754	0.173	0.010	0.086
	(0.526)	(0.348)	(0.204)	(0.270)
Extension contact	−0.056	0.080	−0.024	0.031
	(0.252)	(0.197)	(0.165)	(0.124)
Oxen	−0.154	−0.143	−0.088	−0.031
	(0.118)	(0.111)	(0.061)	(0.092)
Other cattle	0.060**	0.017	0.033**	0.052**
	(0.030)	(0.031)	(0.016)	(0.023)
Small ruminants	0.022***	0.006	0.004	0.002
	(0.008)	(0.007)	(0.006)	(0.006)
Pack animal	−0.079	−0.031	−0.001	−0.048
	(0.066)	(0.069)	(0.037)	(0.045)
Ln(Plot area)	−0.325**	−0.40***	−0.324***	−0.289***
	(0.138)	(0.110)	(0.053)	(0.082)
Soil and water conservation	0.050	−0.190		
	(0.208)	(0.185)		
Reduced tillage			0.234*	−0.063
			(0.126)	(0.154)
Joint significance of zonal dummy variables	8.28**	5.26***	23.60***	6.61*
_cons	−5.484	−1.209	2.585	9.075**
	(8.051)	(5.798)	(3.713)	(4.185)
R-squared	0.542	0.550	0.2965	0.409
Model chi-square	237.087***	10.52***	271.549***	322.617***
Number of observations	131	94	392	202

Note: ***, **, * is significant at 1%, 5%, and 10%, respectively.

TABLE 8-7 Reduced Tillage and Stone Bunds Productivity Analysis Using Switching Regression: Amhara Region

	With Reduced Tillage	Without Reduced Tillage	With Stone Bunds	Without Stone Bunds
Slope	0.010	0.012	0.001	0.022*
	(0.008)	0.013)	(0.009)	(0.012)
Black soil	0.049	−0.077	0.091	0.529***
	(0.164)	(0.196)	(0.122)	(0.203)
Brown soil	0.158	0.192	0.074	0.596***
	(0.163)	(0.210)	(0.126)	(0.195)
Gray soil	0.188	0.249	0.418**	0.407*
	(0.261)	(0.317)	(0.163)	(0.215)
Deep soil	0.048	0.259	0.133	−0.160
	(0.166)	(0.266)	(0.131)	(0.244)
Moderately deep soil	0.103	0.026	−0.086	0.051
	(0.141)	(0.169)	(0.096)	(0.176)
Moderate erosion	−0.111	0.125	0.054	0.011
	(0.128)	(0.194)	(0.106)	(0.142)
Severe erosion	−0.144	0.136	−0.034	−0.238
	(0.192)	(0.242)	(0.209)	(0.218)
Clay soil	−0.049	−0.095	−0.220	−0.308
	(0.187)	(0.258)	(0.176)	(0.314)
Loam soil	0.067	−0.123	−0.054	0.017
	(0.109)	(0.165)	(0.111)	(0.153)
Sandy soil	−0.029	−0.234	−0.173	−0.031
	(0.163)	(0.191)	(0.109)	(0.209)
Highly fertile soil	0.466	0.514	0.680***	0.149
	(0.378)	(0.331)	(0.240)	(0.310)
Medium fertile soil	0.205*	0.152	0.225*	0.211
	(0.122)	(0.161)	(0.117)	(0.161)
Plot distance	0.056	0.085	0.012	0.265
	(0.034)	(0.099)	(0.143)	(0.165)
Rented plot	0.069	0.135	0.322**	0.170
	(0.198)	(0.340)	(0.149)	(0.324)
Irrigation	−0.038	0.736***	0.470***	0.000
	(0.370)	(0.260)	(0.167)	(0.000)
Manure plot	0.079	−0.086	−0.138	0.336
	(0.226)	(0.194)	(0.173)	(0.254)
Ln(plot are)	−0.469***	−0.105	−0.163*	−0.377**
	(0.097)	(0.135)	(0.097)	(0.150)
Market distance	−0.059	0.191***	0.011	−0.088
	(0.051)	(0.069)	(0.047)	(0.074)
Gender	0.431	0.255	−0.046	1.448***
	(0.383)	(0.367)	(0.181)	(0.201)

(continued)

TABLE 8-7 *(Cont.)*

	With Reduced Tillage	Without Reduced Tillage	With Stone Bunds	Without Stone Bunds
Family size	−0.042 (0.029)	0.036 (0.031)	0.001 (0.026)	−0.007 (0.035)
Age	−0.002 (0.007)	−0.007 (0.007)	0.002 (0.005)	−0.004 (0.007)
Education	−0.011 (0.020)	−0.012 (0.026)	0.014 (0.012)	−0.032 (0.024)
Extension contact	0.144 (0.116)	−0.008 (0.152)	0.029 (0.108)	0.001 (0.135)
Population density/km^2	0.002* (0.001)	0.002 (0.002)	−0.052 (0.104)	−0.132 (0.116)
Ln(Rainfall)	0.352 (0.319)	−0.108 (0.264)	0.317** (0.143)	−0.085 (0.238)
Livestock	0.061** (0.031)	0.000 (0.036)	0.072*** (0.028)	−0.008 (0.027)
Farm size	−0.182*** (0.066)	−0.164* (0.091)	−0.339*** (0.085)	−0.114 (0.089)
Soil and water conservation	0.060 (0.102)	0.130 (0.169)		
Fertilizer dummy	0.319** (0.127)	0.740*** (0.212)		
Reduced tillage			0.003 (0.110)	0.264* (0.148)
Joint significant test of zonal dummy variables	37.94***	17.69***	43.82***	62.20***
_cons	3.433 (2.604)	7.037*** (2.132)	4.906*** (1.394)	6.720*** (1.767)
R-squared	0.396	0.468	0.351	0.495
Model chi-square	462.066***	360.041***	1183.943***	474.059***
Number of observations	199	149	234	142

Note: ***, **, * is significant at 1%, 5%, and 10%, respectively.

structures may not be important, but placing appropriate drainage measures could help soil protection during extreme rainfall events.

Blanket recommendations that promote similar conservation measures to all farmers may at best have no effect on productivity or could even be counterproductive. Nuanced promotion of SWC technologies that take into account commonsense environmental realities is therefore likely to be an important determinant of success.

TABLE 8-8 Productivity Impacts from Parametric Regression Analysis

	Reduced Tillage Plots	Non-reduced Tillage Plots	Stone Bund Plots	Non-stone Bund Plots
Amhara region				
Predicted mean gross crop revenue per hectare	1417.405	1304.155	1403.245	1353.066
Predicted mean gross crop revenue difference (standard errors)—ATT†	113.250(93.010)		25.448(180.485)	
Tigray Region				
Predicted mean gross crop revenue per hectare	1880.451	1397.817	1439.777	1226.006
Predicted mean gross crop revenue difference (standard errors)—ATT†	482.634(195.104)**		213.770(84.474)**	

Note: *** significant at 1%; ** significant at 5%; * significant at 10%. †ATT = average treatment effect on the treated.

Notes

1   Amsalu and de Graaff (2007) indeed find that perceived SWC profitability has a statistically significant positive impact on sustained use of those technologies.
2   The rainfall data are long-term averages, spatially interpolated using a climate model (Corbett and White 1998). The minimum and maximum rainfall averaged over the Amhara region for the last fifty years (1953–2003) was 1,303 and 2,457 mm, respectively. Even the minimum average rainfall in Amhara is higher than the maximum annual rainfall (994 mm) of Tigray.
3   These results are consistent with those using kernel and stratification matching, which are available from the author.

References

Amsalu, A., and J. de Graaff. 2007. Sustained Adoption of Soil and Water Conservation Practices in Beressa Watershed, Amhara Region. *Ecological Economics* 61: 294–302.

Becker, O. S., and A. Ichino. 2002. Estimation of Average Treatment Effects Based on Propensity Scores. *The Stata Journal* 2(4): 358–77.

Bekele, W. 2005. Stochastic Dominance Analysis of Soil and Water Conservation in Subsistence Crop Production in the Eastern Ethiopian Highlands: The Case of Hunde-Lafto Area. *Environmental and Resource Economics* 32(4): 533–50.

Benin, S. 2006. Policies and Programs Affecting Land Management Practices, Input Use, and Productivity in the Highlands of Amhara Region, Ethiopia. In *Strategies for Sustainable Land Management in the East African Highlands*, edited by J. Pender, F. Place, and S. Ehui. Washington, DC: International Food Policy Research Institute, 217–256.

Byiringiro, F., and T. Reardon. 1996. Farm Productivity in Rwanda: Effects of Farm Size, Erosion, and Soil Conservation Investments. *Agricultural Economics* 15: 127–36.

Corbett, J. D. and J.W White. 1998. Using the sub-annual climate models for meso-resolution spatial analyses, Temple, USA, Mud Springs Geographers, Inc.

Holden, S. T., B. Shiferaw, and J. Pender. 2001. Market Imperfections and Profitability of Land Use in the Ethiopian Highlands: A Comparison of Selection Models with Heteroskedasticity. *Journal of Agricultural Economics* 52(2): 53–70.

Kaliba, A. R. M., and T. Rabele. 2004. Impact of Adopting Soil Conservation Practices on Wheat Yield in Lesotho. In: Bationo, A., Ed., Managing Nutrient Cycles to Sustain Soil Fertility in Sub-Saharan Africa. Tropical Soil Biology and Fertility Institute of CIAT

Kassie, M., and S. T. Holden. 2006. Parametric and Non-parametric Estimation of Soil Conservation Adoption Impact on Yield. Paper presented at the International Association of Agricultural Economists Conference. August 2006, Gold Coast, Australia.

Pender, J., and B. Gebremedhin. 2007. Determinants of Agricultural and Land Management Practices and Impacts on Crop Production and Household Income in the Highlands of Tigray, Ethiopia. *Journal of African Economies* 17(3): 395–450.

Shiferaw, B., and S. T. Holden, 2001. Farm-level benefits to investments for mitigating land degradation: empirical evidence from Ethiopia. Environment and Development Economics 6, 335–358.

Shively, G. E. 1998a. Modeling Impacts of Soil Conservation on Productivity and Yield Variability: Evidence from a Heteroskedastic Switching Regression. Paper presented at annual meeting of the American Agricultural Economics Association. August 1998, Salt Lake City, Utah.

———. 1998b. Impact of Contour Hedgerow on Upland Maize Yields in the Philippines. *Agroforestry Systems* 39: 59–71.

———. 1999. Risks and Returns from Soil Conservation: Evidence from Low-income Farms in the Philippines. *Environmental Monitoring and Assessment* 62: 55–69.

CHAPTER 9

Soil Conservation and Small-scale Food Production in Highland Ethiopia: A Stochastic Metafrontier Approach

HAILESELASSIE MEDHIN AND GUNNAR KÖHLIN

A s is true throughout East Africa and indeed much of the developing world, agriculture is the most important economic activity in Ethiopia, providing livelihoods for more than three-fourths of the population and half the gross domestic product. The bulk of agricultural output comes from subsistence farmers in the highlands, which are home to more than 80% of Ethiopia's population (World Bank 2004).

These mountainous areas are characterized by high dependency on rainfall, traditional technology, and high population pressure, all with one of the lowest average productivity levels in the world. According to World Bank (2005), the 2002–2004 average yield was 1,318 kg/hectare, which is less than 60% of other low-income countries and 40% of the world average. There are only three tractors per arable 100 square km, which is again far less than the 66 tractors for low-income countries. As a result, agricultural value-added per Ethiopian worker is $123 versus $375 for low-income countries and $776 for the world. One reason for this low productivity is very serious land degradation (Hurni 1988). According to Swinton and Gebremedhin (2003), over 10 million hectares will not support cultivation by 2010. The Ethiopian Highland Reclamation Study reports that 50% of the highlands is significantly and 25% seriously eroded; over 2 million hectares were uncultivable as early as 1985 (Tegegn 1995).

As discussed throughout this book and elsewhere in the literature, soil and water conservation (SWC) is one of the most important ways to improve agricultural productivity in areas with high land degradation and limited access to modern inputs.[1] As with any farm technology, SWC adoption depends on the nature of the circumstances each farmer faces. Most studies, including the chapters in this book, stress that farmers consider a variety of issues. These factors include risk and time preference (Shiferaw and Holden 1999; Shively 2001; Yesuf 2004), land tenure (Alemu 1999; Swinton and Gebremedhin 2003), off-farm activities and resource endowments (Grepperud 1995; Shively 2001), yield variability (Shively 1999), public policies and market structure (Holden et al. 2001; Diagna 2003; Yesuf and Pender 2005).

The economics literature on the productivity-enhancing effects of SWC is inconclusive, however. Using Ugandan plot-level data Byiringaro and Reardon (1996) find that farms with more SWC investment have much better land productivity. Nyangena (2006), after controlling for plot quality, concludes that SWC increases yields of degraded plots in Kenya. On the other side, Kassie in Chapter 8 and earlier work (e.g., Kassie 2005) finds that—depending on circumstances—conserved plots may have lower returns than non-conserved plots. Holden et al. (2001) use data from an Ethiopian highland village and also find that conservation technologies have no significant positive effect on land productivity. Shively (1999) assesses the effect of hedgerow contours relative to conventional tillage practices in the Philippines. He finds that although hedgerows can increase average yield they also increase variability.

This chapter aims to contribute to the literature on SWC by decomposing productivity into technology and technical efficiency (TE) effects (Coelli et al. 1998). A firm is said to be technically inefficient if it produces less output from a given input bundle than the maximum output that can be attained from the input bundle at the current level of technology. If there is improvement in technology, the maximum output attainable from the input bundle increases. Hence there is positive technology effect, and where no decrease in TE occurs because of the new technology, productivity increases. If there is a decrease in technical efficiency as a result of the new technology, the productivity change will depend on the magnitude of the decline in TE.

Let us illustrate this using a simple diagrammatic example shown in Figure 9-1. Suppose a new farm technology T is introduced to a community. Assume the production function before the introduction of T is F, and the production function with T is F^T. *Hence there is a positive technology effect because F^T is above F.* For each technology, a *TE* production level should lie *on* the production function and a technically inefficient one *below* it. Suppose a farmer in a community endowed with input bundle X is fully technically efficient under the old technology and hence operates at point Q and produces output Y^Q. Where would the point of production be using the new technology input bundle X? If the farmer continues to be fully efficient with the new technology, s/he will operate at point T and produce Y^T. Hence, productivity will increase. If *TE* decreases with application of the new technology—or if the production point is below T—the impact on productivity will depend on the new point in comparison to Q.

If the new point is above Q, but below T there is a decline in *TE*, but there is still improvement in productivity because the technology effect dominates.

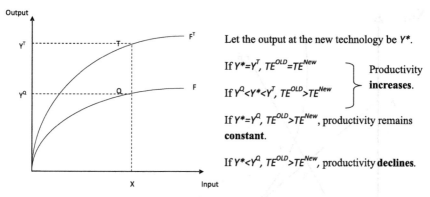

Let the output at the new technology be Y^*.

If $Y^*=Y^T$, $TE^{OLD}=TE^{New}$ $\Big\}$ Productivity

If $Y^Q<Y^*<Y^T$, $TE^{OLD}>TE^{New}$ $\Big\}$ **increases**.

If $Y^*=Y^Q$, $TE^{OLD}>TE^{New}$, productivity remains **constant**.

If $Y^*<Y^Q$, $TE^{OLD}>TE^{New}$, productivity **declines**.

FIGURE 9-1 Improvement in Technology and Technical Efficiency

That is, even if the farmer produces a lower output from X compared to the maximum in the new technology, he or she still manages to produce more output from X than with the old technology. If the new point is below or at P, the decline in *TE* is so high that productivity decreases or stays constant.[2] In the real world farmers are most likely be technically inefficient even at the old technology (producing below Q). It is easy to locate the point under the new technology where $TE^{OLD} = TE^{NEW}$ and repeat the same analysis. It is therefore possible for *TE* to increase, but the key point is to measure *TE* at each production point with reference to the maximum output attainable with the applicable technology.

Conceptual Framework

This section has two main goals. The first is to introduce the economic theory and measurement of efficiency. Next, we discuss stochastic metafrontier analysis and its application to the efficiency of production agents with varying technologies. Our eventual aim is to assess how such efficiency measurement can be applied to SWC and aid policymakers and development agents in matching SWC technologies and agro-economic environments.

Figure 9-2 represents efficiency using conventional isoquant and isocost diagrams based on Farrell (1957). Assume a firm produces output Y using two inputs X_1 and X_2. SS' is a set of fully efficient combinations of X_1 and X_2 producing output Y^*—an isoquant. Similarly, AA' is a minimum cost input-price ratio or isocost. Now assume that the actual input combination point to produce Y^* is P. Clearly, the firm is experiencing both technical and allocative inefficiencies. The measures can be calculated as follows:

$$TE = \frac{OQ}{OP} = 1 - \frac{QP}{OP}. \tag{9.1}$$

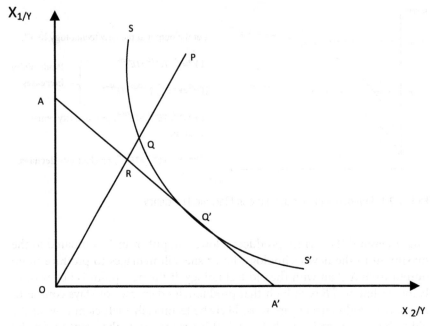

FIGURE 9-2 Input-oriented Technical and Allocative Efficiencies

Similarly, allocative efficiency (AE) is defined as:

$$AE = \frac{OR}{OQ}. \qquad (9.2)$$

Equation (9.2) suggests the possible reduction in costs that can be achieved by using correct input proportions such that the isocost line is tangent to the isoquant. Note that it is possible for a technically efficient point to be allocatively inefficient, though an allocatively efficient point is technically efficient. The total economic efficiency (EE) is defined as the product of the two measures, TE and AE (Coelli et al. 1998):

$$EE = \left(\frac{OQ}{OP}\right) \bullet \left(\frac{OR}{OQ}\right) = \frac{OR}{OP}. \qquad (9.3)$$

The above measure requires that the production function be estimated. In this chapter we use a stochastic metafrontier approach. Stochastic metafrontier analysis incorporates regional and technological differences across firms[3] and is an umbrella of stochastic production frontiers estimated for groups of firms operating under different technologies. The stochastic frontier was first introduced by Aigner et al. (1977) and relaxes the assumptions that the production frontier is common to all firms and that inter-firm variation in performance is attributable only to differences in technical efficiency. In general terms a stochastic production frontier can be written as:

$$Y_i = f\left(X_i;\beta\right)e^{\left(V_i - U_i\right)} \quad i = 1, 2, \ldots, n_j , \qquad (9.4)$$

where Y_i = output of the ith firm, X_i = vector of inputs, β = vector of parameters, V_i = random error term, and U_i = inefficiency term.

In agricultural analysis V_i captures random factors like measurement error, weather, and drought. V_i is assumed to be an independently and identically distributed normal random variable with constant variance independent of U_i, which is assumed to be a non-negative exponential truncated variable of $N(\mu_i, \sigma^2)$, where μ_i is defined by some inefficiency model (Coelli et al. 1998; Battese and Rao 2002). Most econometric frontiers assume one-sided inefficiency disturbances (Førsund et al. 1980).

Another important point is the choice of f ($\cdot$). Battese (1992) notes that translog or Cobb-Douglas production functions are the most commonly used functional forms and we use the Cobb-Douglas specification. For a sample of n_j firms the jth group making up the stochastic frontier is defined by:

$$Y_{ij} = f\left(X_{ij};\beta\right)e^{\left(V_{ij} - U_{ij}\right)}, \quad i = 1, 2, \ldots, n_j . \qquad (9.5)$$

Assuming the production function is Cobb-Douglas or translog, this can be rewritten as:

$$Y_{ij} = f\left(X_{ij};\beta\right)e^{\left(V_{ij} - U_{ij}\right)} = e^{X_{ij}\beta + V_{ij} - U_{ij}} \quad i = 1, 2, \ldots, n_j . \qquad (9.6)$$

And the "overall" stochastic frontier is:

$$Y_i = f\left(X_i;\beta^*\right)e^{\left(V_i^* - U_i^*\right)} = e^{X_i\beta^* + V_i^* - U_i^*} \quad i = 1, 2, \ldots, n; \ \ n = \sum n_j . \qquad (9.7)$$

Equation (9.7) is the stochastic metafrontier function. The superscripts * differentiate the parameters and error terms of the metafrontier function from the group-level stochastic functions. Note that Y_i and X_i remain the same, but separate output and input samples of different groups are pooled. The metafrontier is an envelope of the stochastic frontiers of the groups; we therefore can have two TE estimates for firms, one with respect to the frontier of its group and another with respect to the metafrontier.

The parameters of the group and metafrontiers can be estimated using maximum likelihood. After estimating β and β^*, it is expected that the deterministic values $X_{ij}\beta$ and $X_i\beta^*$ should satisfy the inequality $X_{ij}\beta \le X_i\beta^*$, because $X_i\beta^*$ is from the metafrontier. According to Battese and Rao (2002), this relationship can be written as:

$$1 = \frac{e^{X_{ij}\beta}}{e^{X_i\beta^*}} \cdot \frac{e^{Vi}}{e^{Vi^*}} \cdot \frac{e^{-Ui}}{e^{-Ui^*}}. \qquad (9.8)$$

Equation (9.8) indicates that differences between estimated parameters of a given group and the metafrontier should arise from technology gap (TGR), random error (RER), and technical efficiency (TER) ratios. Equation (9.9) presents the details.

$$TGR_i = \frac{e^{Xij\beta}}{e^{Xij\beta^*}} \equiv e^{-Xi(\beta^* - \beta)}, \quad RER_i = \frac{e^{Vi}}{e^{Vi^*}} \equiv e^{Vi - Vi^*} \quad \text{and}$$

$$TER_i = \frac{e^{-Ui}}{e^{-Ui^*}} \equiv \frac{TE_i}{TE_i^*}. \tag{9.9}$$

The technology gap ratio shows the technology gap for each group in terms of the best industry technology, assuming all groups have access. The TGR and the TER can be estimated for each firm. Note that $0 < TE_i \leq 1$ and $0 < TE_i^* \leq 1$, $TE_i^* \leq TE_i$ and $TER \geq 1$. The random error ratio is not observable, because it is based on the non-observable disturbance term V_i. Therefore, for estimation purposes equation (9.8) can be rewritten as:

$$1 = \frac{e^{Xij\beta}}{e^{Xij\beta^*}} \cdot \frac{e^{-Ui}}{e^{-Ui^*}} = TGR_i \times TER_i. \tag{9.10}$$

Combining (9.9) and (9.10) gives:

$$TE_i^* = TE_i \times TGR_i. \tag{9.11}$$

From (9.11) we see that the *TE* relative to the metafrontier function is the product of the *TE* relative to the group frontier and the *TGR* of the technology group. This is a very important identity in that it enables us to estimate the extent to which the productivity of a firm or group of firms could be increased if it adopted the best available technology. In our case we use this approach to estimate the technology gap between plots with and without SWC and investigate the role of different soil conservation practices in defining the technology of farm plots.

The metafrontier curve is an envelope of stochastic frontier curves of the technology groups. If each technology group has at least one firm using the best industry technology (i.e., if one firm's TGR = 1), the metafrontier connects these best-practice firms from all groups. In cases like for groups 2 and 4 in Figure 9-3 where no firm in a group uses best technology, the stochastic frontier of the group lies below the metafrontier.

According to Battese et al. (2004), for a Cobb-Douglas production function (or any function log-linear in parameters) the metafrontier can be estimated using a simple optimization.

$$\text{Minimize } X'\beta$$

$$\text{Subject to } X_i\beta \leq X_i\beta^* \tag{9.12}$$

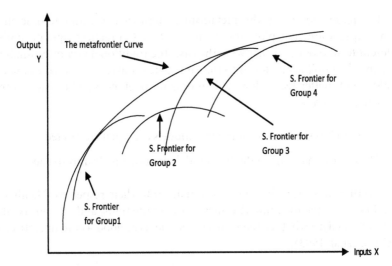

FIGURE 9-3 The Stochastic Metafrontier Curve

In Equation (9.12), X' is the row vector of means of all inputs for each technology group; β is the vector group coefficients, and β^* is the vector of meta-coefficients we seek. This is simply a linear programming problem, and the βs are the maximum likelihood coefficients of the group stochastic frontier from FRONTIER 4.1[4] estimations. The constraint inequality represents the assumption that an envelope exists. Once we obtain the solutions to our linear programming problem (β^*s) it is easy to calculate the $TGRs$ and metafrontier technical efficiencies. From equation (9.9) we know that $TGR_i = \dfrac{e^{Xij\beta}}{e^{Xi\beta^*}}$, and from equation (9.11) $TE_i^* = TE_i \times TGR_i$, with TE_i estimated in our group stochastic frontiers.

Data and Empirical Specification

The study uses the Ethiopian Environmental Household Survey data collected by the Departments of Economics at Addis Ababa University and University of Gothenburg and managed by the Environmental Economics Policy Forum for Ethiopia (EEPFE). The survey covers six *weredas*[5] in two highland zones in the Amhara Regional State in northern Ethiopia. Given the similarity of the survey areas to other highland regions, we believe that the results can be used to comment on policies that aim to increase the productivity of agriculture throughout highland Ethiopia. We focus on two major crop types (teff[6] and wheat) and analyze the use of stone bund terracing, soil bund terracing, and bench terracing SWC technologies, which are some of the most important used in Ethiopia and throughout East Africa; plots with other SWC technologies are also included in estimating the metafrontier.

The reason for using the metafrontier approach to estimate efficiency is the expectation that plots under different SWC practices operate under different technologies. If that were the case, the traditional way of estimating efficiency by pooling all plots into the same data set may give biased estimates as plots with better technology will appear more efficient. This leads to the following hypotheses:

H_0: The estimates from the two models are equal (restricted).

H_1: The estimates from the two models are not equal (unrestricted).

Substituting maximum likelihood estimates into their respective likelihood functions gives the maximized values of these functions, and we can use the LR test to analyze whether the restricted model is as good as the unrestricted (Griffiths et al. 1993).

$$\lambda_{LR} = -2[L(H_0) - L(H_1)]$$

where λ_{LR} is the likelihood ratio statistic, $L(H_0)$ is the maximized value of the log-likelihood function under H_0 (restricted log-likelihood function), and $L(H_1)$ is the maximized value of the log-likelihood under H_1 (unrestricted log-likelihood function). λ_{LR} has an approximate $\chi^2_{(j)}$ distribution, where j is the number of restrictions under H_0. We reject H_0 when $\lambda_{LR} > \chi^2_c$, where χ^2_c is a chosen critical value from the χ^2_c distribution. The pooled specification restricts parameter estimates in the separate estimation to be equal to those of the pooled estimation. If H_0 is not rejected it means the pooled stochastic frontier represents the data. The number of these restrictions is the degrees of freedom.

Plots are grouped according to the type of SWC technologies applied, and we use seven SWC groups: (1) *no SWC* used; (2) *soil bunds*; (3) *bench terraces*; (4) *stone bunds*; (5) *contour furrowing*; (6) *contour plowing* and (7) *others*. We then estimate the production function parameters (farm input coefficients) and technical efficiency for each SWC technology group. Descriptive statistics of farm inputs are presented in Table 9-1.

TABLE 9-1 SWC Technology Groups

Variable	Mean Value				
	None	*Soil bunds*	*Stone bunds*	*Bench terraces*	*Pooled*
Yield (kg/ha)	1035.87	815.32	955.76	943.79	1076.02
Labor (days)	38.61	39.80	48.97	47.96	43.42
Traction (days)	5.74	5.23	4.79	6.99	6.15
Fertilizer (ETB*)	24.99	14.16	20.66	28.74	30.57
Manure (kg)	41.79	37.26	58.03	64.96	57.70

*ETB = Ethiopian *birr*

Some input variables like fertilizer and manure have zero values for some plots. As the model requires values to be in logarithms, dummy variables that detect such values are included in the production function. Our stochastic frontier functions also include fertilizer and manure dummies. The mathematical expressions of the stochastic frontier to be estimated are:

$$\ln Output_{ij} = \beta_{oj} \ln land_{ij} + \beta_{oj} \ln labor_{ij} + \beta_{oj} \ln traction_{ij} + \beta_{oj} \ln seed_{ij}$$
$$+ \beta_{oj} \ln fert_{ij} + \beta_{oj} \ln man_{ij} + \beta_{oj} FertD_{ij} + \beta_{oj} manD_{ij} + \exp(V_{ij} - U_{ij}). \qquad (9.13)$$

For each technology group (9.13) is simultaneously estimated using FRONTIER 4.1. In addition to the β coefficients, the TE of each plot and the log-likelihood functions are estimated. The pooled model is also estimated in the same manner. The pooled estimation is critical to conducting the LR test and forming the metafrontier. If a metafrontier is justified, the next step is to solve the linear programming problem using Mathematica 5.1, estimate the metacoefficients, and calculate the TGR and metafrontier efficiency of each plot in each technology group. We then assess the extent of the technology-based productivity differences between the first four SWC groups that were displayed in Table 9-1.

Discussion of Results

Table 9-2 presents the maximum likelihood coefficient and TE estimates of the stochastic production function for the technology groups, including the pooled data. Most estimates are positive and significantly different from zero, which is in accord with expectations. We also find that plots cultivated under all SWC technologies experience technical inefficiency, and those with no SWC technologies are least efficient.

As our likelihood ratio test statistic (λ) is 371.24 ($\alpha = 0.005$) we reject the hypothesis that the pooled stochastic estimation is a correct representation of the data, which implies there are significant technology differentials across the four SWC groups. The parameter estimates of the pooled data are still relevant for the production function, however, because technological variability within the pooled representation only biases the TE estimates and productivity comparisons are only possible if we know the position of group frontiers relative to the metafrontier, which is the best technology frontier for all groups.

Aside from technology differentials, we also must address non-random forces that affect output. For example, some studies have shown that SWC investment takes away land and labor from production (Shiferaw and Holden 1998, 2001). The land input does not account for land and labor inputs that are lost to SWC.[7] Hence, output lost due to land used for SWC structures is treated as if it were technical inefficiency.

TABLE 9-2 Coefficients of the Production Function

Variable	Coefficient (t–ratio)				
	None	*Soil bunds*	*Stone bunds*	*Bench terraces*	*Pooled*
Constant	4.2487**	4.3130**	4.5430**	5.8685**	4.3618**
	(20.4663)	(4.7752)	(14.7057)	(6.3162)	(35.4124)
Land	0.3496**	0.3436*	0.2377**	0.7310**	0.3149**
	(8.0103)	(1.6600)	(4.3778)	(3.8998)	(12.0299)
Labor	0.2794**	0.2992	0.1408**	0.0082	0.2290**
	(5.9290)	(1.6263)	(2.5056)	(0.04735)	(8.4113)
Traction	0.2081**	−0.1521	0.3395**	0.3116**	0.2071**
	(5.3726)	(−1.1041)	(6.0643)	(2.4046)	(8.2998)
Seed	0.2502**	0.2678**	0.1193**	0.0866	0.2337**
	(10.3058)	(2.8847)	(3.1099)	(1.2983)	(15.5857)
Fertilizer	−0.0878	−0.1332	−0.0248	0.0381	−0.0038
	(−1.5757)	(−0.6145)	(−0.1993)	(0.3887)	(0.1303)
Manure	0.0613	0.1215	0.1541**	−0.1461	0.0235
	(1.1711)	(0.6437)	(2.3833)	(−1.0644)	(0.8208)
Fertilizer use	0.4230	0.5533	0.0575	−0.0605	0.0410
dummy	(1.6245)	(0.5705)	(0.0973)	(−0.1352)	(0.2832)
Manure use	−0.2959	−0.3893	−0.6386*	1.1504	0.0116
dummy	(−1.1053)	(−0.3986)	(−1.8279)	(1.4711)	(0.0758)
Mean TE	0.65498	0.77971	0.67615	0.68731	0.67825

**statistically significant at 5% level; *statistically significant at 10% level

Plots with steeper slopes are also less productive than those with less slope. As shown in Table 9-3, the null hypothesis that conserved and non-conserved plots have similar slopes is rejected at a very high level of significance, with steeper plots having a higher likelihood of being conserved. The null hypothesis that conserved and unconserved plots have similar soil qualities is also rejected at the 8% significance level, suggesting that conserved plots have systematically lower quality soils. SWC could therefore be productivity enhancing while still not leading to highest yields due to plot characteristics.

TABLE 9-3 Mann-Whitney Test for Plot Slope and Soil Quality

Plot characteristics	Hypotheses	Z-value	P-value
Slope	H_0: Slope^C = Slope^U	6.6104***	0.000
	H_1: Slope^C > Slope^U		
Soil quality	H_0: Soil Q^C = Soil Q^U	1.4220*	0.0778
	H_1: Soil Q^C < Soil Q^U		

C = conserved; U = unconserved; *** significant at α = 0.01; * significant at α = 0.1

TABLE 9-4 Yield and Input Use Comparison

Variable	With soil conservation		Without soil conservation		Difference	
	Mean	*Std. dev.*	*Mean*	*Std. dev.*	*With-without*	*P-value*
Yield (kg/h)	928.97	834.91	1035.87	981.43	−106.90	0. 0421
Labor (days)	47.18	39.98	38.62	30.10	8.56	0.0002
Manure (kg)	55.71	177.61	41.80	305.40	13.91	0.3309
Fertilizer (ETB)	21.05	62.71	24.99	77.27	−3.94	0.3594
Traction (oxen days)	5.28	4.91	5.74	5.78	−0.46	0.1391
Seed (kg)	18.05	40.74	26.19	40.18	−8.14	0.0000

Std. dev. = standard deviation; ETB = Ethiopian *birr*; US$1 = ETB 8.5.

The existence of these confounding factors means that, while higher TE indicates that chosen inputs are used more effectively on conserved plots, they may not generate higher yields. As shown in Table 9-4, the average yield for plots with conservation is significantly lower than for plots without conservation. This is consistent with the results of the previous chapter, which also analyzed SLM investments in Amhara Regional State. Conserved plots also use more labor and less seed, but differences in manure, fertilizer and traction are insignificant.

Table 9-5 presents the results of the metafrontier estimation and compares with the group estimates from Table 9-4. The mean TGR quantifies the average gap between group technology and overall technology. Group frontiers with TGR values of 1.00 are tangent to the metafrontier, and we find that all group frontiers are tangent to the metafrontier except bench terraces.[8] Plots with soil bunds have the lowest mean TGR of 0.78, which means that even if all soil

TABLE 9-5 Technology Gaps and Metafrontier TE

Technology Group	Variable	Mean
None	TGR[a]	0.9494
	Meta TE	0.62061
	Group TE	0.65497
Stone bunds	TGR	0.9539
	Meta TE	0.64607
	Group TE	0.67614
Soil bunds	TGR	0.7806
	Meta TE	0.60600
	Group TE	0.77970
Bench terraces	TGR	0.9629
	Meta TE	0.65748
	Group TE	0.68733

[a]TGR = technology gap ratio

TABLE 9-6   SWC and Frontier (Best-practice) Plots

	Total Number of Plots Cultivated under This SWC Technology	% Share	Total Number of Frontier Plots Cultivated under This SWC Technology	% Share
None	667	54.3	75	51.2
Stone Bunds	357	29.1	56	38.1

bund plots use the best available group technology they will still have output about 22% lower than if they used the best technology of the whole sample.

So are SWC technologies useful? From Table 9-5 we see that plots without SWC have no significant technology gap relative to plots with SWC technologies; soil bunds have the highest technology gap (lowest TGR). From this one might conclude that SWC is not terribly useful, but one must recall that SWC technologies tend to be adopted on plots with poor natural conditions. It is therefore quite possible that SWC improves the composite technology of conserved plots, but unconserved plots still have better technologies on average; using an SWC technology is still helpful if conserved plots would have performed worse had they not been conserved or if unconserved plots could have performed better had they been conserved.

One additional way to shed light on the role of SWC in improving agricultural productivity is to study plots with TGRs very close to 1.000. Table 9-6 presents an assessment of the impact of SWC on agricultural productivity for the 147 of 1,228 plots that use best practice in the four technology groups, including plots without conservation (54.3% of the sample).

The results show that the share of plots with stone bunds in the pool of frontier plots (38.1%) is significantly higher than their share in the total pool (29.1%), but there is a lower percentage of plots with no SWC in the frontier pool than in the overall sample. A comparison of plots with different slopes shows that the proportion of very steep plots in the best-practice plots pool is double the overall sample; advantaged plots with better soil and topographic conditions, with or without SWC, therefore define the best technology, but SWC appears to help disadvantaged plots reach the frontier. One can perform similar analyses of factors like slope, soil quality, moisture abundance, rainfall, soil type, and market conditions and look for best combinations of SWC technologies and agro-economic factors. Indeed, the stochastic metafrontier approach can be used to match SWC technologies with the external conditions in which they work best.

Concluding Remarks

Most SWC technologies are promoted only after being tested for positive technology effects, but it is impossible to test the efficiency of SWC in experimental labs. Sometimes even in the field SWC fails to deliver visible on-farm productivity improvements. Indeed, in small-scale agriculture the prevalence of heterogeneities is usually the biggest challenge in the design of strategies to boost productivity. We find that yields on plots with SWC technologies are lower than on those without conservation. It is, of course, possible that technologies are promoted and adopted in improper circumstances. For example, as discussed in Chapter 8, SWC success depends on agro-ecological factors like climate and topography; it therefore may fail because it is adopted in the wrong place.

The group stochastic frontier estimates show that plots with SWC are relatively more efficient than plots without soil conservation, but this does not explain why yields on unconserved plots are greater than conserved plots. The stochastic metafrontier estimation suggests that a key explanation is that plots with different SWC practices operate with different technologies. Efficiency estimates therefore could give biased results.

An in-depth look at the characteristics of best-practice plots shows that SWC is a key part of plot technologies, but plot characteristics also weigh in; plots may therefore not be appropriately matched. One way to match plots appropriately is to identify best-practice farms using the stochastic metafrontier approach and then combine SWC technologies, climate, soil, topography, and market characteristics that offer highest levels of efficiency. Once we take these steps, it is found that SWC is likely to help disadvantaged plots remain productive.

Acknowledgments

The authors would like to acknowledge Sida (the Swedish International Development Cooperation Agency) through the Environment for Development (EfD) initiative at University of Gothenburg, Sweden, for financial support. The authors are also grateful for access to data collected by the departments of economics at Addis Ababa University and the University of Gothenburg through a collaborative research project titled "Strengthening Ethiopian Research Capacity in Resource and Environmental Economics," financed by Sida/SAREC.

Notes

1   Nyangena (2006) also notes that inorganic fertilizers could have negative environmental externalities if not properly used.
2   For example, a new SWC technology could be associated with specific organizational capabilities that the farmer does not possess.
3   The stochastic metafrontier applied here is mainly adopted from Battese and Rao (2002) and Battese et al. (2004).
4   FRONTIER 4.1 is software commonly used to estimate production frontiers and efficiency.
5   *Wereda* is the name for the second lowest administrative level in Ethiopia. The lowest is *kebele*.

6   Teff is a tiny grain that is used to make *injera*, Ethiopia's most common food. The grain has many variants.
7   The labor cost of SWC does not affect the TE estimate even though it could affect allocative efficiency (AE) estimates.
8   Looking at the TGR data for each plot in each SWC technology shows that not even a single plot under bench terraces has a TGR = 1.

References

Aigner, D. J., C. A. K. Lovell, and P. Schmidt. 1977. Formulation and Estimation of Stochastic Frontier Production Function Models. *Journal of Econometrics* 6: 21–37.

Alemu, T. 1999. Land Tenure and Soil Conservation: Evidence from Ethiopia. Ph.D. thesis No. 92. Department of Economics, University of Gothenburg, Sweden.

Battese, G. E. 1992. Frontier Production Functions and Technical Efficiency: A Survey of Empirical Applications in Agricultural Economics. *Agricultural Economics* 7: 185–208.

Battese, G. E., and D. S. P. Rao. 2002. Technology Gap, Efficiency, and a Stochastic Metafrontier Function. *International Journal of Business and Economics* 1(2): 1–7.

Battese, G. E., D. S. P. Rao, and C. J. O'Donnell. 2004. A Metafrontier Production Function for Estimation of Technical Efficiencies and Technology Gaps for Firms Operating under Different Technologies. *Journal of Productivity Analysis* 21: 91–103.

Byiringiro, F., and T. Reardon. 1996. Farm Productivity in Rwanda: Effects of Farm Size, Erosion, and Soil Conservation Investment. *Agricultural Economics* 15: 127–36.

Coelli, T., D. S. P. Rao, C. O'Donnell, and G. E. Battese. 1998. *An Introduction to Productivity and Efficiency and Productivity Analysis*, 2nd ed. New York: Springer.

Diagana, B. 2003. Land Degradation in Sub-Saharan Africa: What Explains the Widespread Adoption of Unsustainable Farming Practices? Draft Working Paper. Bozeman, MT: Department of Agricultural Economics and Economics, Montana State University.

Farrell, M. J. 1957. The Measurement of Productive Efficiency. *Journal of Royal Statistical Society, Series A* 120: 253–81.

Førsund, F. R., C. A. K. Lovell, and P. Schmidt. 1980. A Survey of Frontier Production Functions and Their Relationship to Efficiency Measurement. *Journal of Econometrics* 13: 5–25.

Grepperud, S. 1995. Soil Conservation and Government Policies in Tropical Areas: Does Aid Worsen the Incentives for Arresting Erosion? *Agricultural Economics* 12: 120–40.

Griffiths, W.E., R.C. Hill, and G.G. Judge. 1993. *Learning and Practicing Econometrics*. Hoboken, NJ: John Wiley & Sons, Inc.

Holden, S., B. Shiferaw, and J. Pender. 2001. Market Imperfections and Land Productivity in Ethiopian Highlands. *Journal of Agricultural Economics* 52(3): 53–70.

Hurni, H. 1988. Degradation and Conservation of the Resources in the Ethiopian Highlands. *Mountain Research and Development* 8(2/3): 123–30.

Kassie, M. 2005. Technology Adoption, Land Rental Contracts, and Agricultural Productivity. Ph.D. dissertation (2005: 20). Norwegian University of Life Sciences, Ås, Norway.

Nyangena, W. 2006. Essays on Soil Conservation, Social Capital, and Technology Adoption. Ph.D. thesis No. 148, Department of Economics, University of Gothenburg, Sweden.

Shiferaw, B., and S. Holden. 1998. Resource Degradation and Adoption of Land Conservation Technologies in the Ethiopian Highlands: A Case Study in Andit Tid, North Shewa. *Agricultural Economics* 18: 233–47.

———. 1999. Soil Erosion and Smallholders' Conservation Decisions in the Highlands of Ethiopia. *World Development* 27(4): 739–52.

———. 2001. Farm-level Benefits to Investments for Mitigating Land Degradation: Empirical Evidence for Ethiopia. *Environment and Development Economics* 6: 336–59.

Shively, G. E. 1999. Risks and Returns from Soil Conservation: Evidence from Low-income Farms in the Philippines. *Agricultural Economics* 21: 53–67.

———. 2001. Poverty, Consumption Risk, and Soil Conservation. *Journal of Development Economics* 65: 267–90.

Swinton, S. M., and B. Gebremedhin. 2003. Investment in Soil Conservation in Northern Ethiopia: The Role of Land Tenure Security and Public Programs. *Agricultural Economics* 29: 69–84.

Tegegn, G. 1995. An Assessment of Ethiopia's Agricultural Land Resources. In *Ethiopian Agriculture: Problems of Transformation, Proceedings of the 4th Annual Conference on the Ethiopian Economy*, edited by D. Aredo and M. Demeke. AAU Press.

World Bank. 2004. *World Development Report 2004: Making Services Work for Poor People*. Washington, DC: World Bank.

———. 2005. *World Development Indicators 2005*. Washington, DC: The World Bank.

Yesuf, M. 2004. Risk, Time, and Land Management under Market Imperfections: Application to Ethiopia. Ph.D. thesis No. 139. Department of Economics, University of Gothenburg, Sweden.

Yesuf, M., and J. Pender. 2005. Determinants and Impacts of Land Management Technologies in the Ethiopian Highlands: A Literature Review. Draft paper. Addis Ababa, Ethiopia: Environmental Economics Policy Forum in Ethiopia (EEPFE) and Washington, DC: International Food Policy Research Institute (IFPRI). www.efdinitiative.org/research/projects/project-repository/economic-sector-work-on-poverty-and-land-degradation-in-ethiopia (accessed on January 26, 2011).

Public Policies and Sustainable Land Management Investments

PART IV

Public Policies and Sustainable Land Management Investments

Policy Instruments to Reduce Downstream Externalities of Soil Erosion and Surface Runoff

ANDERS EKBOM

P revious chapters have discussed many of the factors affecting investments in soil and water conservation. The purpose of this chapter is to analyze how potential policy instruments spur these investments and address downstream externalities from soil loss and agricultural runoff, with a particular focus on information, regulation, strengthening property rights, and market based instruments. Identifying appropriate instruments is particularly important for small-scale agriculture, which is often characterized by erosive soils, dispersed non-point pollution, and asymmetric information between polluters and downstream victims; these settings are especially common in mountainous tropical developing countries (Kerr 2002; Landell-Mills and Porras 2002; Pagiola et al. 2005; Wunder 2005).

Although some economic research has addressed downstream externalities of upland farming (e.g., Clark et al. 1985; Moore and MacCarl 1987; Ribaudo 1986; de Janvry et al. 1995), this phenomenon has mainly been studied by other scientists (Morris and Fan 1998; Rowan et al. 2001). We particularly focus on hilly areas in tropical developing countries where small-scale farming is practiced on steep slopes using mainly family labor, and we rely on the experience of Kenya to guide our discussion. The chapter is organized as follows: The next section characterizes downstream externalities. We then discuss policy instruments to prevent or mitigate externalities and conclude by summarizing key policy conclusions.

Downstream Externalities and the Economics of Soil Management

Total economic costs of downstream externalities due to erosion and agricultural runoff are estimated to be substantial. For instance, Palmieri et al. (2003) report that reservoir sedimentation is a major problem for most of the world's

45,000 large (>15m tall) manmade dams; global costs of sedimentation are on the order of $13 billion per year. This is equivalent to 45 km^3 in lost annual water storage capacity. Magrath and Arens (1989) estimate the off-site social costs of soil erosion in Java, Indonesia, to be $35 to $91 million per year. Smith (1992) reports that the annual addition to stock externalities[1] from U.S. agriculture is 5.3% and flow externalities represent 4.6% of agricultural output value.

Eroded soil and agricultural runoff carry pathogens like viruses and bacteria, as well as suspended particles and nitrates, and over one billion people depend on unprotected wells, rivers, lakes, and reservoirs for their water. All these contaminants increase human morbidity and mortality downstream.[2] Unsafe water causes more than 130 million people to suffer from severe intestinal parasites,[3] viral infections, and bacterial diseases (Bartram et al. 2005). There are also potentially important ecological effects. Nitrogen runoff, phosphorus, and other nutrients from chemical fertilizers and soils increase the incidence of eutrophication and algae blooms (Matson et al. 1997; Ayoub 1999), which can contain potent toxins[4] that negatively affect shellfish, aquaculture (Shumway 1990), fish populations, and marine ecosystems (Anderson 1995; Horner et al. 1997). Nutrient leaching may also facilitate the spread of invasive species like the water hyacinth in Lake Victoria (Naidu et al. 1998; Bartram and Chorus 1999). Additional flow externalities include the formation of scours and gullies that increase the frequency and impact of floods. Flooded areas constitute breeding grounds for malaria mosquitoes and other vectors.

Coral reefs are complex ecosystems, which are especially sensitive to stress factors like nutrient supply, sedimentation, and temperature changes. They play critical roles for fisheries, tourism, and as sources of income for poor people in developing countries (Andersson 2004), but approximately 22% of all coral reefs are under high or medium threats from inland pollution and soil erosion (Bryant et al. 1998). Empirical studies on reefs include work on Japan (Shimoda et al. 1998), the Philippines (Hodgson 1990), the Great Barrier Reef in Australia (Fabricius and De'ath 2001), Guam in the Pacific Ocean (Golbuu et al. 2003), Venezuela (Bastidas et al. 1999), Indonesia (Holmes et al. 2000), the U.S. Virgin Islands (Hubbard 1986), Puerto Rico (Loya 1976), Hawaii (Stimson and Larned 2000), Barbados (Tomascik and Sander 1987), Costa Rica (Cortes and Risk 1985), Brazil (Costa Jr. et al. 2000), and Kenya (McClanahan and Obura 1997; Van Katwijk et al. 1993).

Soil erosion and surface runoff accumulate, causing so-called stock externalities such as sedimentation of coastal environments, reservoirs, irrigation canals, and other water supply structures. Such stock externalities have been documented as major environmental issues in Spain (De Vente et al. 2004), Tunisia (Hamed et al. 2002), Nepal (Ross and Gilbert 1999), United Kingdom (White et al. 1997), Mexico (Fernex et al. 2001), Dominican Republic (Nagle 2002), and Kenya (Saenyi and Chemelil 2003).

Thomas (1994) reports that sediment yields from five major catchments in Ethiopia, Kenya, Tanzania, South Africa, and Lesotho range between 290 and 1,980 tons/km^2/year.[5] The Masinga Reservoir on the Tana River in Kenya regulates water for five hydropower installations that produce 453 MW per year. The total capacity is 2,345 million m^3, but in 1988 the sediment yield in the Masinga Reservoir was estimated at 12.8 million tons/year (Palmieri et al. 2003). Stoorvogel and Smaling (1990, 1998) report substantial nutrient losses in Eastern and Southern Africa, with the average in the Kenyan highlands estimated at 73 kg/ha of nitrogen (N), 7 kg/ha of phosphorus (P), and 51 kg/ha of potassium (K).[6]

Sedimentation of reservoirs generates direct costs, defensive expenditures, and forgone profits. Direct costs include damages to turbine blades and clogging of access pipes. Defensive expenditures include afforestation, conservation of watersheds, dredging of water reservoirs, ports, and irrigation channels, and upstream construction of diversion channels and weirs. Reduced dam service affects profits and electricity production capacity. On average, sedimentation reduces global reservoir capacity by 1% per year (Mahmood 1987). The Tarbela Dam Reservoir in Pakistan, for example, has an average annual sediment inflow of 200 million tons; the Sefid-Rud Dam in Iran lost 63% of its capacity after 17 years, and about the same was true for the Ichari Dam in India after 6 years (Palmieri et al. 2001). Cruz et al. (1988) conclude that erosion in the Philippines reduced the life of two water reservoirs 35% to 40%.

The economics of soil management discussed throughout this volume dates back at least to Wilcox (1938) and Bunce (1942). Significant modern treatments include papers by Burt (1981), McConnell (1983), Shortle and Miranowski (1987), Barbier (1990), Barrett (1991), Clarke (1992), LaFrance (1992), Goetz (1997), Grepperud (1996, 1997a, 1997b, 2000), Smith et al. (2000), Yesuf (2004), and Ekbom (2007). A key focus of this literature is that soil is capital that must be managed if agricultural production is to be maintained at high levels. As is true for any type of capital, investments are required.[7]

A farmer facing the potential loss of soil capital has the choice of depleting soil or conserving using tilling, crop choice, and conservation measures like terraces, bunds, cut-off drains, and hedgerows. It is not obvious that a farmer "must" conserve soil, and a privately optimal investment strategy may include some soil erosion. Objective functions of farmers and policymakers may also differ, because upstream farmers are unlikely to fully include externalities in

TABLE 10-1 Reservoirs and Storage Loss in Kenya's Central Highlands

Reservoir	Commis- sioned	Main purpose	Catchment area (km^2)	Original capacity (10^6 m^3)	Current capacity (10^6 m^3)	Storage loss (%)
Masinga	1981	Hydro-power	7335	1560	1100	29%
Ruiru	1950	Water supply	67	2.98	2.496	16%

Source: Palmieri et al. (2003)

158 Anders Ekbom

their decisions or even be aware of the effects downstream. We emphasize that a downstream area can be upstream in a larger geographic or hydrologic context.

Policy Instruments to Mitigate Downstream Effects

One approach a policymaker who is concerned with off-site externalities caused by poor upland farmers could use is to identify combinations of policy instruments that maximize the discounted social net benefit from agricultural production. Policymakers need to consider on- and off-site effects, but they must distinguish among them, because each has different impacts, actors, and economic incentives. Indeed, farmers may have incentives to prevent on-site, but not off-site effects.

Below we use a theoretical model to examine the on- and off-farm effects of policy instruments to address key market failures in developing country hill agriculture. These include direct regulation, information and extension, charges and fees, improved property rights, and payments for environmental services.[8] Farmers use too little soil conservation labor from a social point of view, because most farmers do not internalize the downstream benefits of soil conservation. The discrepancy in conservation labor supply is illustrated in Figure 10-1, where the privately optimal labor supply $(CL_{PRIVATE})$ is chosen where the *private* marginal benefit of conservation labor equals the market wage rate (w). As opposed to the privately rational farmer, a social planner needs to acknowledge that on-farm soil conservation produces downstream (social) benefits in terms of reduced siltation, water purification, hydrological balance, etc. Figure 10-1 illustrates that socially optimal labor supply is greater than the privately optimal supply. Moving the privately optimal labor supply towards the socially optimal curve presupposes a farmer who is willing to produce public goods and supply conservation labor up to the point where the *socially* marginal benefit of conservation labor meets the wage rate. This level represents more intensive soil conservation (corresponding to CL_{SOCIAL}) compared with the privately optimal supply (represented by $CL_{PRIVATE}$).

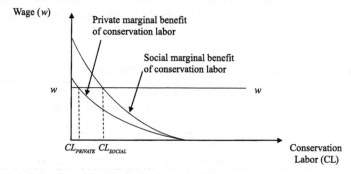

FIGURE 10-1 Privately and Socially Optimal Supply of Conservation Labor

Direct Regulation

Under direct regulation farmers are obliged to supply cultivation and/or conservation inputs. Governments have frequently used policies like bans on certain types of cultivation and soil conservation requirements to address soil erosion and runoff (Hudson 1985; Morgan 1986). In Kenya, for instance, starting in 1937 and continuing until independence in 1963, soil conservation was compulsory on cultivated land. Mandatory engineering solutions like construction of labor-intensive bench terraces, cut-off drains, stone gabions, and retention ditches were also prescribed (Kimaru 1998). Although choice and implementation of policies have changed considerably since independence, direct regulation is still an important element of Kenya's conservation policy: farmers are still required to conserve soil. Conservation officers keep records of soil conservation measures, and failure to adopt may subject farmers to sanctions. There are also bans—frequently violated—on cultivating plots with greater than 60% slope and along riverbanks. Vertical plowing is also prohibited.

The regulatory approach to soil conservation has largely been unsuccessful, because incentives for soil conservation are insufficient.[9] As explained in Ekbom (2007) and illustrated above, labor and cash-constrained farmers who cultivate erosion-prone soils on steep slopes and cause soil loss and fertilizer runoff have few incentives to sufficiently prevent downstream externalities; when private benefits are smaller than social returns, only a share of conservation benefits accrue privately.

Information and Extension

Outreach to farmers has been used in East Africa to promote sustainable agriculture and has largely replaced earlier land-use policies based on coercion. Information can be a cost-effective policy instrument (Sterner 2003), and in Kenya it is considered rather successful (OPTO 2006; Kimaru 1998; Lundgren 1993).[10] However, downstream damages remain very serious problems, suggesting that information is a useful—but insufficient—policy instrument to optimally prevent damages. For example, information can be very effective for addressing on-farm productivity effects, but because of poor incentives it may do little to reduce off-farm externalities.

Charges and Fees

In principle, the external costs to downstream victims could be internalized using a fee on degrading inputs or practices. As explained in Ekbom (2007) and illustrated in Figure 10-2, a rational farmer with secure land rights would use agricultural labor such that the value of the marginal product (VMP) equals the market wage (w) plus the marginal effect on the farmers' own soil capital (A). This corresponds to $AL_{PRIVATE}$. However, since cultivation causes

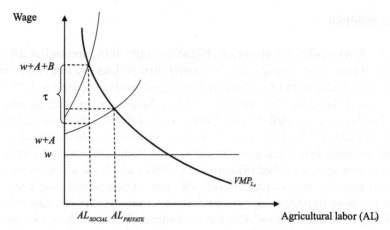

FIGURE 10-2 The Effect of an Erosion Charge on Agricultural Cultivation Labor

downstream externalities (represented by B) it is optimal to reduce the use of agricultural labor to AL_{SOCIAL}.

If one introduces a charge (τ) that corresponds to the cost of downstream externalities, these costs could be internalized. However, since small-scale farming is characterized by non-point source pollution, such measures are hard to enforce. Charges on downstream damages are also politically sensitive. Farmers may be rich and powerful or, as in many tropical countries, so poor that they can hardly pay additional fees. In principle it might be possible to construct a package in which charges are counteracted by reducing other fees, but erosion charges would likely still be difficult to implement because of monitoring and enforcement problems.

Improved Property Rights

Land tenure security affects both investment incentives and the availability of resources to finance investments (Feder and Feeny 1991). Farmers holding title to their land may use it as collateral, which facilitates credit and enables investments like terracing and tree plantations. Feder and Onchan (1987), for example, find that ownership security is positively correlated with land-improving investments; in northern Ethiopia this appears to also be true for soil conservation investments (Shiferaw and Holden 2000; Alemu 1999).

In Kenya's central highlands land tenure security is relatively high compared to neighboring countries. A majority of smallholders hold title to surveyed, registered, and adjudicated lands, which can be bought and sold on the open market and used as collateral. A problem, though, is that land in practice is owned or controlled by men and inherited only by sons. Women who head households, divorced women, or widows enjoy no real rights to land or formal titles. Consequently, they have few incentives to invest. An important reform is therefore to adjust the current institutions governing land owner-ship to facilitate registration and strengthen women's rights to land.

Strengthening tenure security on private farmland is a necessary condi-tion, but it will be insufficient to fully prevent downstream externalities. Complementary measures would be to strengthen the rights of downstream water users to clean water and to provide economic incentives for intensified soil conservation. Such steps might include increasing support for soil conser-vation, decontamination of existing water sources, redistribution among existing users, and increasing freshwater supply.

A critical question is whether downstream water users are always enti-tled to clean water or if upstream farmers hold the right to pollute. It seems natural to argue that all downstream victims should be compensated, but in Kenya and many other developing countries, upland farming existed long before downstream hydropower production, irrigation, and coastal tourism. Farmers can therefore, perhaps, legitimately claim a historical right, though in some areas lowland farmers settled earlier than those in uplands (Ochieng and Maxon 1992).

Payments for Environmental Services

Payments for environmental services (PES) compensate farmers for producing public environmental services[11] and offer positive incentives for desirable actions. Figure 10-3 presents the labor model amended to include PES. We know that a competitive farmer would apply soil conservation labor that corresponds to $CL_{PRIVATE}$. However, at this level downstream effects occur that are not internalized. The privately optimal level of soil conservation is therefore too little to sufficiently prevent downstream externalities.

To encourage farmers to move toward the socially optimal level of conser-vation labor supply (CL_{SOCIAL}) some form of compensation is one option. In our case this corresponds exactly to s and may be provided as a cash payment for the environmental services farmers produce. Historically, in Kenya compensation

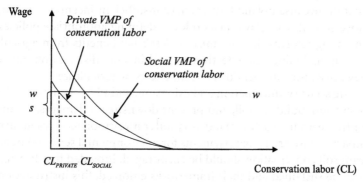

FIGURE 10-3 Conservation Labor Supply and the Effect of PES in the Form of a Wage

has been provided to prevent private yield losses. For example, Kenya's government has provided subsidies for tree seedlings, tools, and implements, as well as cash payments to encourage farmers to conserve soil.

PES is a bit different, because it is used to encourage soil conservation and reduce downstream externalities (Pagiola and Platais 2002; Pagiola 2008; Kerr 2002; Landell-Mills and Porras 2002; Gutman 2003; Pagiola et al. 2005; Wunder 2005). PES establishes property rights but presupposes a broader environmental and social perspective on soil conservation than merely agricultural yields (FAO 2007). In practice, PES is usually paid for forgone benefits. In our case this would mean farmers who conserve soil are compensated for lost agricultural output or direct costs of conservation.

There are, of course, opportunity costs when PES payments are made (Landell-Mills and Porras 2002; Pagiola and Platais 2002; Pagiola et al. 2005). PES is therefore appropriate when: (a) transaction costs are low among the ecosystem service providers, downstream beneficiaries, and mediators (e.g., the government); (b) PES payments to farmers equal or exceed the costs of providing environmental services; and (c) the benefits to downstream beneficiaries equal or exceed opportunity costs. A key practical issue is setting the appropriate payment level when marginal benefits and costs of soil conservation labor are unknown and optimal conservation levels are therefore uncertain.

Other critical issues in implementing PES include characterization of the ecological services, establishment of sustainable financing mechanisms, design and implementation of effective payment systems, and establishment of adequate institutional frameworks (Campos et al. 2005). The design of PES institutional frameworks is critical to ensuring cost-efficiency, effective monitoring, and enforcement to prevent free riding.

Summary and Conclusions

Agricultural production commonly causes downstream damage due to soil erosion and nutrient runoff. This substantial problem requires government intervention because upland farmers have insufficient incentives to conserve soil. Governments can play crucial roles in defining appropriate policies and implementing reforms that optimize society's net benefits from agricultural production, including cost-effectiveness and income distribution outcomes, but they must also consider downstream effects of agriculture.

Tenure security, information and extension, and PES schemes are likely to reduce erosion, build up soil, and prevent downstream damages. Command and control measures such as mandatory soil conservation have potential negative impacts, so we argue that payments for environmental services targeted at upstream soil conservation should be encouraged. Provided that PES institutions can be sustained and their frameworks enforced, this instrument offers incentives to build up soil capital and reduce downstream externalities.

Acknowledgments

Helpful comments from Gardner Brown, Thomas Sterner, Ed Barbier, E. Somanathan, Carolyn Fischer, Peter Parks, Francisco Alpizar, and my fellow contributors to this book are gratefully acknowledged, as is funding for the research from Sida (Swedish International Development Cooperation Agency) through the Environmental Economics Unit at the University of Gothenburg.

Notes

1   External costs pertain to freshwater and marine recreation, water storage, navigation, flooding, irrigation, commercial fishing, municipal water treatment, and municipal and industrial use.

2   Nitrates are converted in the digestive tracts into toxic nitrite. Nitrite causes the "blue baby syndrome" (methemoglobinemia), which impairs the blood's ability to transport oxygen within the body. This syndrome is particularly common among infants and may cause death (Younes and Bartram 2001).

3   Including helminths like roundworm (*Ascaris*), whipworm (*Trichuris*), and hookworm (*Necator/Ancylostoma*).

4   For example, cyanobacteria (bluegreen algae) and dinoflagellates. Andersson (1995) presents a summary of major harmful or toxic algal species.

5   Equivalent to 2.9–19.8 tons/ha/year

6   For comparison, mean fertilizer use in sub-Saharan African agriculture amounts to 11 kg per harvested hectare. Developing countries apply on average 62 kg per hectare (FAO 1995).

7   Soil capital consists of a range of biological, physical, and chemical properties, including macronutrients (e.g., nitrogen, phosphorus, potassium), micronutrients (e.g., copper), cat-ion exchange capacity, moisture, permeability, structure, clay-sand-silt content, and pH-level, as well as pathogens from cattle and human excreta, pesticides, and manure.

8   Due to the complexities and substantial institutional requirements associated with use of other policy instruments such as tradable permits, they are not considered in this chapter (see Sterner 2003).

9   The theoretical model and technical details on which the reasoning in this section rests is presented in Ekbom (2007).

10   The positive effects of extension advice have been contested by Evenson and Mwabu (2001) and Gautam and Anderson (1999), who found limited evidence of positive effects on productivity.

11   Examples of environmental services might include protecting freshwater quality, controlling hydrological flows, reducing suspension and sedimentation of water systems, preventing floods and landslides, conserving biodiversity, and sequestering carbon.

References

Alemu, T. 1999. Land Tenure and Soil Conservation: Evidence from Ethiopia. Ph.D. thesis, Department of Economics, University of Gothenburg, Sweden.

Anderson, D. M. 1995. Toxic Red Tides and Harmful Algal Blooms: A Practical Challenge in Coastal Oceanography. *Reviews of Geophysics* 33(S1): 1189–1200.

Andersson, J. 2004. Welfare, Environment, and Tourism in Developing Countries. Ph.D. thesis, Department of Economics, University of Gothenburg, Sweden.

Ayoub, A.T. 1999. Fertilizers and the Environment. *Nutrient Cycling in Agro-Ecosystems* 55(2): 117–21.

Barbier, E. 1990. The Farm-level Economics of Soil Conservation: The Uplands of Java. *Land Economics* 66(2): 199–211.

Bartram, J., and I. Chorus, eds. 1999. *Toxic Cyanobacteria in Water: A Guide to their Public Health Consequences, Monitoring, and Management.* World Health Organization. www.who.int/water_sanitation_health/resourcesquality/toxcyanobacteria. pdf (accessed January 26, 2011)

Bartram J., K. Lewis, R. Lenton and A. Wright. 2005. Millennium Project: Focusing on Improved Water and Sanitation for Health. *The Lancet* 365: 810–12.

Bastidas, C., D. Bone, and E. M. García. 1999. Sedimentation Rates and Metal Content of Sediments in a Venezuelan Coral Reef. *Marine Pollution Bulletin* 38(1): 16–24.

Barrett, S. 1991. Optimal Soil Conservation and the Reform of Agricultural Pricing Policies. *Journal of Development Economics* 36(2): 167–87.

Bryant, D. G., L. Burke, J. McManus, and M. Spalding. 1998. *Reefs at Risk: A Map-based Indicator of Threats to the World's Coral Reefs.* Washington, DC: World Resources Institute.

Bunce, A. 1942. *The Economics of Soil Conservation.* Ames: Iowa State College Press.

Burt, O. R. 1981. Farm Level Economics of Soil Conservation in the Palouse Area of the Northwest. *American Journal of Agricultural Economics* 63(1): 83–92.

Campos, J. J., F. Alpizar, B. Louman, and J. Parotta 2005. *An Integrated Approach to Forest Ecosystem Services.* In *Forests in the Global Balance—Changing Paradigms,* edited by G. Mery, R. Alfaro, M. Kanninen, and M. Lobovikov. IUFRO World Series Volume 17. Helsinki.

Clark, E. H., J. A. Haverkamp, and W. Chapman. 1985. *Eroding Soils: The Off-farm Impacts.* Washington, DC: The Conservation Foundation.

Clarke, H.R. 1992. The Supply of Non-degraded Agricultural Land. *Australian Journal of Agricultural Economics* 36(1): 31–56.

Cortes J. N., and M. J. Risk. 1985. A Reef under Siltation Stress: Cahuita, Costa Rica. *Bulletin of Marine Science* 36: 339–56.

Costa Jr., O. S., Z. M. Leao, M. Nimmo, and M. J. Attrill. 2000. Nutrification Impacts on Coral Reefs from Northern Bahia, Brazil. *Hydrobiologia* 440: 370–415.

Cruz, W., H. A. Francisco, and Z. T. Conway. 1988. The On-site and Downstream Costs of Soil Erosion in the Magat and Pantabangan Watersheds. *Journal of Philippine Development* 15 (261): 85–111.

de Janvry, A., E. Sadoulet, and B. Santos. 1995. Project Evaluation for Sustainable Development: Plan Sierra in the Dominican Republic. *Journal of Environmental Economics and Management* 28(2): 135–54.

De Vente, J., J. Poesen, and G. Verstraeten. 2004. The Application of Semi-quantitative Methods and Reservoir Sedimentation Rates for the Prediction of Basin Sediment Yield in Spain. *Journal of Hydrology* 305(October): 1–24.

Ekbom, A. 2007. Economic Analysis of Agricultural Production, Soil Capital, and Land Use in Kenya. Ph.D. thesis, Department of Economics, University of Gothenburg, Sweden.

Evenson, R. E., and G. Mwabu. 2001. The Effect of Agricultural Extension on Farm Yields in Kenya. *African Development Review* 13(1): 1–23.

Fabricius, K. E., and G. De'ath. 2001. Environmental Factors Associated with the Spatial Distribution of Crustose Coralline Algae on the Great Barrier Reef. *Coral Reefs* 19: 303–09.

FAO (United Nations Food and Agriculture Organization). 1995. *World Agriculture: Towards 2010.* Edited by N. Alexandratos. Rome: FAO.

———. 2007. *The State of Food and Agriculture 2007: Paying Farmers for Environmental Services.* Rome: FAO.

Feder, G., and D. Feeny. 1991. Land Tenure and Property Rights: Theory and Implications for Development Policy. *World Bank Economic Review* 5(1): 135–53.

Feder, G., and T. Onchan. 1987. Land Ownership Security and Farm Investment in Thailand. *American Journal of Agricultural Economics* 69(2): 311–20.

Fernex, F., P. Zárate-del Valle, H. Ramírez-Sánchez, F. Michaud, C. Parron, J. Dalmasso, G. Barci-Funel, and M. Guzman-Arroyo. 2001. Sedimentation Rates in Lake Chapala (Western Mexico): Possible Active Tectonic Control. *Chemistry Geology* 177(3–4): 213–28.

Gautam, M., and J. R. Anderson. 1999. *Reconsidering the Evidence on the Returns to T&V Extension in Kenya.* Policy Research Working Paper No 2098. Washington, DC: Operations Evaluations Department, the World Bank.

Goetz, R. U. 1997. Diversification in Agricultural Production: A Dynamic Model of Optimal Cropping to Manage Soil Erosion. *American Journal of Agricultural Economics* 79(2): 341–56.

Golbuu, Y., S. Victor, E. Wolanski, and R. Richmond. 2003. Trapping of Fine Sediment in a Semi-enclosed Bay, Palau, Micronesia. *Estuarine, Coastal, and Shelf Science* 57(5–6): 941–49.

Grepperud, S. 1996. Population Pressure and Land Degradation: The Case of Ethiopia. *Journal of Environmental Economics and Management* 30(1): 18–33.

———. 1997a. Soil Conservation as an Investment in Land. *Journal of Development Economics* 54(2): 455–67.

———. 1997b. Poverty, Land Degradation, and Climatic Uncertainty. *Oxford Economic Papers* 49(4): 586–608.

———. 2000. Optimal Soil Depletion with Output and Price Uncertainty. *Journal of Environment and Development Economics* 5(3): 221–40.

Gutman, P. 2003. *From Goodwill to Payments for Environmental Services: A Survey of Financing Options for Sustainable Natural Resource Management in Developing Countries.* Washington, DC: WWF Macroeconomics Program Office.

Hamed, Y., J. Albergel, Y. Pépin, J. Asseline, S. Nasri, P. Zante, R. Berndtsson, M. El-Niazy, and M. Balah. 2002. Comparison between Rainfall Simulator Erosion and Observed Reservoir Sedimentation in an Erosion-sensitive Semi-arid Catchment. *CATENA* 50(1): 1–16.

Hodgson, G. 1990. Sediment and the Settlement of Larvae of the Reef Coral Pocillopora Damicornis. *Coral Reefs* 9: 41–43.

Holmes, K. E., E. N. Edinger, Hariyadi, G. V. Limmon, and M. J. Risk. 2000. Bioerosion of Live Massive Corals and Branching Coral Rubble on Indonesian Coral Reefs. *Marine Pollution Bulletin* 7: 606–17.

Horner, R. A., D. L. Garrison, and F. G. Plumley. 1997. Harmful Algal Blooms and Red Tide Problems on the U.S. West Coast. *Limnology and Oceanography* 42(5): 1076–88.

Hubbard, D. K. 1986. Sedimentation as a Control of Reef Development: St. Croix, U.S.V.I. *Coral Reefs* 5: 117–25.

Hudson, N. 1985. *Soil Conservation*. Batsford, UK: Batsford Academic and Educational Publishers.

Kerr, J. 2002. Watershed Development, Environmental Services, and Poverty Alleviation in India. *World Development* 30(8): 1387–1400.

Kimaru, G. 1998. *Kenya's National Soil and Water Conservation Programme (NSWCP)*. Copenhagen, Denmark: Department for Development Research, Danish Institute for International Studies.

LaFrance, J. T. 1992. Do Increased Commodity Prices Lead to More or Less Soil Degradation? *Australian Journal of Agricultural Economics* 36(1): 57–82.

Landell-Mills, N., and I. T. Porras. 2002. *Silver Bullet or Fool's Gold? A Global Review of Markets for Forest Environmental Services and their Impact on the Poor, Instruments for Sustainable Private Sector Forestry*. London: International Institute for Environment and Development.

Loya, Y. 1976. Effects of Water Turbidity and Sedimentation on the Community Structure of Puerto Rican Corals. *Bulletin of Marine Science* 26: 450–66.

Lundgren, L. 1993. *Twenty Years of Soil Conservation in Eastern Africa*. Regional Soil Conservation Unit Report No. 9. Nairobi, Kenya: Majestic Printing Works.

Magrath, W., and P. Arens. 1989. The Costs of Soil Erosion on Java: A Natural Resource Accounting Approach. Environment Department Working Paper No. 18. Washington, DC: The World Bank.

Mahmood, K. 1987. Reservoir Sedimentation: Impact, Extent, and Mitigation. World Bank Technical Paper. Washington, DC: The World Bank.

Matson, P. A., W. J. Parton, A. G. Power, and M. J. Swift. 1997. Agricultural Intensification and Ecosystem Properties. *Science* 277(5325): 504–09.

McClanahan, T. R., and D. Obura. 1997. Sedimentation Effects on Shallow Coral Communities in Kenya. *Journal of Experimental Marine Biology and Ecology* 209(1–2): 103–22.

McConnell, K. E. 1983. An Economic Model of Soil Conservation. *American Journal of Agricultural Economics* 65(1): 83–89.

Moore, W. B., and B. A. MacCarl. 1987. Off-site Costs of Soil Erosion: A Case Study in the Willamette Valley. *Western Journal of Agricultural Economics* 12 (July): 42–49.

Morgan, R. P. C. 1986. *Soil Erosion and Conservation*. New York: John Wiley & Sons, Inc.

Morris, G. L., and J. Fan. 1998. *Reservoir Sedimentation Handbook: Design and Management of Dams, Reservoirs, and Watersheds for Sustainable Use*. New York: McGraw Hill.

Nagle, G. N. 2002. The Contribution of Agricultural Erosion to Reservoir Sedimentation in the Dominican Republic. *Water Policy* 3(6): 491–505.

Naidu R., S. Baskaran, and R. S. Kookana. 1998. Pesticide Fate and Behavior in Australian Soils in Relation to Contamination and Management of Soil and Water: A Review. *Australian Journal of Soil Research* 36(5): 715–64.

Ochieng, W. R., and R. M. Maxon, eds. 1992. *An Economic History of Kenya.* Nairobi, Kenya: East African Educational Publishers Ltd.

OPTO (OPTO International Development Consultancy). 2006. Impact Assessment of the National Agriculture and Livestock Extension Programme (NALEP), Phase 1. OPTO International AB, http://aideffectivenesskenya.org/index.php?option=com_docman&task=doc_download&gid=6338&Itemid=376 (accessed on January 26, 2011).

Pagiola, S. 2008. Payments for Environmental Services in Costa Rica. *Ecological Economics* 65(4): 712–24.

Pagiola, S., and G. Platais. 2002. *Payments for Environmental Services.* Washington, DC: World Bank/IBRD.

Pagiola, S., A. Arcenas, and G. Platais. 2005. Can Payments for Environmental Services Help Reduce Poverty? An Exploration of the Issues and the Evidence to Date. *World Development* 33(2): 237–53.

Palmieri, A., S. F. Farhed, and A. Dinar. 2001. Economics of Reservoir Sedimentation and Sustainable Management of Dams. *Journal of Environmental Management* 61(2): 149–63.

Palmieri, A., S. F. Farhed., G. W. Annandale, and A. Dinar. 2003. *Reservoir Conservation: Economic and Engineering Evaluation of Alternative Strategies for Managing Sedimentation in Storage Reservoirs; The RESCON Approach Vol. I.* Washington, DC: The World Bank.

Ribaudo, M. O. 1986. Consideration of Off-site Impacts in Targeting Soil Conservation Programs. *Land Economics* 62(Nov.): 402–11.

Ross, J., and R. Gilbert. 1999. Lacustrine Sedimentation in a Monsoon Environment: the Record from Phewa Tal, Middle Mountain Region of Nepal. *Geomorphology* 27(3–4): 307–23.

Rowan, J. S., L. E. Price, C. P. Fawcett, and P. C. Young. 2001. Reconstructing Historic Reservoir Sedimentation Rates Using Data-based Mechanistic Modelling. *Physics and Chemistry of the Earth* 26(1): 77–82.

Saenyi, W. W., and M. C. Chemelil. 2003. Modelling of Suspended Sediment Discharge for Masinga Catchment Reservoir in Kenya. *Journal of Civil Engineering* 8: 89–98.

Shiferaw, B., and S. T. Holden. 2000. Policy Instruments for Sustainable Land Management: the Case of Highland Smallholders in Ethiopia. *Agricultural Economics* 22: 217–32.

Shimoda, T., T. Ichikawa, and Y. Matsukawa. 1998. Nutrient Conditions and Their Effects on Coral Growth in Reefs around Ryukyu Islands. *Bulletin of the National Research Institute of Fisheries Science* 12: 71–80.

Shortle, J. S., and J. A. Miranowski. 1987. Intertemporal Soil Resource Use: Is it Socially Excessive? *Journal of Environmental Economics and Management* 14(June): 99–111.

Shumway, S. E. 1990. A Review of the Effects of Algal Blooms on Shellfish and Aquaculture. *Journal of the World Aquaculture Society* 21: 65–104.

Smith, E. G., M. Lerohl, T. Messele, and H. H. Janzen. 2000. Soil Quality Attribute Time Paths: Optimal Levels and Values. *Journal of Agricultural and Resource Economics* 25(1): 307–24.

Smith, V. K. 1992. Environmental Costing for Agriculture: Will It Be Standard Fare in the Farm Bill of 2000? *American Journal of Agricultural Economics* 74(5): 1076–88.

Sterner, T. 2003. *Policy Instruments for Environmental and Natural Resource Management.* Washington, DC: Resources for the Future.

Stimson J., and S. T. Larned. 2000. Nitrogen Efflux from Sediments of a Subtropical bay and the Potential Contribution to Macro-algal Nutrient Requirements. *Journal of Experimental Marine Biology and Ecology* 252: 159–80.

Stoorvogel, J. J., and E. M. A. Smaling. 1990. *Assessment of Soil Nutrient Depletion in Sub-Saharan Africa: 1983–2000,* Vol. 1: Main Report. Report No. 28. Wageningen, The Netherlands: Winand Staring Center for Integrated Land, Soil, and Water Research.

———. 1998. Research on Soil Fertility Decline in Tropical Environments: Integration of Spatial Scales. *Nutrient Cycling in Agro-ecosystems* 50: 151–58.

Thomas, M. F. 1994. *Geomorphology in the Tropics: A study of Weathering and Denudation in Low Latitudes.* Chichester, UK: John Wiley and Sons, Ltd.

Tomascik, T., and F. Sander. 1987. Effects of Eutrophication on Reef-building Corals: Structure of Scleractinian Coral Communities on Fringing Reefs, Barbados, West Indies. *Marine Biology* 94: 53–75.

Van Katwijk, M., N. Meier, R. van Loon, E. van Hove, W. Giesen, G. van der Velde, and D. den Hartog. 1993. Sabaki River Sediment Load and Coral Stress: Correlation between Sediments and Condition of the Malindi Watamu Reefs in Kenya (Indian Ocean). *Marine Biology* 117: 675–83.

White, P., D. P. Butcher, and J. C. Labadz. 1997. Reservoir Sedimentation and Catchment Sediment Yield in the Strines Catchment, U.K. *Physics and Chemistry of the Earth* 22(3–4): 321–28.

Wilcox, W. W. 1938. Economic Aspects of Soil Conservation. *Journal of Political Economy* 46(5): 702–13.

Wunder, S. 2005. *Payments for Environmental Services: Some Nuts and Bolts.* Jakarta, Indonesia: Center for International Forestry Research (CIFOR).

Yesuf, M. 2004. Risk, Time, and Land Management under Market Imperfections: Applications to Ethiopia. Ph.D. thesis, Department of Economics, University of Gothenburg, Sweden.

Younes M., and J. Bartram. 2001. Waterborne Health Risks and the WHO Perspective. *International Journal of Hygiene and Environmental Health* 204(4): 255–63.

CHAPTER 11

Incentives for Sustainable Land Management in East African Countries

BERHANU GEBREMEDHIN

Confronted with increasing populations on already degraded lands, worsening poverty, and declining food production per capita, many developing countries continue to struggle with poverty reduction and natural resource sustainability. As has been discussed throughout this volume increasing agricultural production is an important component in that struggle. With a shrinking frontier, increases in agricultural production need to come from improvements in productivity (Eicher 1994), but significant increases in agricultural productivity cannot be attained if the land resource base is being degraded.

As we have already seen, decline in the productive capacity of land, is a major problem confronting many East African countries and is likely a key constraint on agricultural growth. The highlands of Ethiopia, Kenya, Tanzania, and Uganda have strong agricultural potential, but since the early 1970s they have experienced severe land degradation, including soil erosion, nutrient depletion, overgrazing, deforestation, and soil moisture stress (Jones 2002; Mbaga-Semgalawe and Folmer 2000; Gebremedhin 1998; Stahl 1993; Zake 1992).

A number of initiatives for sustainable land management (SLM) have been undertaken in these countries, especially since the early 1970s. Many involve incentives for small farmers to invest in land conservation. However, the results of these efforts have been mixed. In this chapter I discuss the nature and types of incentives used for SLM in East Africa and evaluate their efficacy. The paper concludes with suggestions for improving the performance of these instruments.

Causes of Land Degradation

The proximate causes of land degradation are many. They include cultivation of steep slopes and erosion-prone soils, low vegetative cover, burning of dung and crop residues, feeding crop residues to animals, reducing fallow periods, deforestation, and overgrazing. The underlying causes of these practices may include population growth, poverty, high costs for fertilizer, fuel, and animal feed, limited access to fertilizers, limited farmer knowledge, insecure land tenure, and credit market imperfections. Because the proximate causes are the consequences of inappropriate land management conditioned on the underlying factors, instruments for SLM must primarily address underlying causes; unless real causes are addressed, sustainable land management is unlikely.

In order to examine the relationship between incentives and land degradation, it is necessary to lay out the connection between the underlying factors and land degradation. I will focus here on land tenure security, product and factor market development, and extension service provision.

As was discussed in by Nyangena in Chapter 3 and Kububo-Mariara and Linderhof in Chapter 4, tenure security affects the extent to which farmers benefit from investments in land. It is reasonable that if the risk of losing rights to land is high, farmers would be less likely to invest. While land tenure security does not necessarily require private ownership or title, the rights of users must be recognized and respected over a sufficient period to allow for secure investments. Though we often think of tenure security with regard to cultivated lands, the importance of tenure security also applies to communal forest and grazing lands. Without effective collective action for communal land management, externalities and free rider problems could result in degradation and lack of investment. Appropriate specification and enforcement of property rights over communal lands is therefore essential for SLM.

When factor and product markets are underdeveloped or not competitive, farmers receive poor market signals, often resulting in inefficient decisions. For example, as several chapters have emphasized, farmers may not be able to make investments in land management if credit markets are poorly developed or nonexistent; improving credit market performance may therefore allow farmers to invest in sustainable practices. Agricultural extension and technical assistance services can enhance farmers' capacities to conserve land. However, top-down and unresponsive extension service may offer little help to farmers. In some cases, top-down extension service may even displace local efforts, doing more harm than good (Gebremedhin et al. 2003b).

The underlying causes of land degradation are more amenable to the use of policy instruments like macroeconomic, sectoral, population, land tenure, and use rights policies. They may also be better addressed by agricultural research and extension, input supply, and credit, market, and infrastructure development programs than are proximate causes. For example, infrastructure and market development may improve the profitability of farming, thus encouraging investment in sustainable land management practices. Better supply of animal feed could reduce the use of crop residues for that purpose. Credit programs reduce liquidity constraints, and improved tenure security increases returns—both potentially encouraging investment and reducing soil mining.

Why Incentives?

For several decades the problem of land degradation was largely considered a physical problem to be solved through technology (Sanders and Cahill 1999). Hence, research focused largely on technical solutions and providing extension services to promote those technologies; land users often had little input. When such approaches failed, researchers, development practitioners, and policymakers were forced to recognize that land degradation is a consequence of underlying economic, social, and institutional factors. As a result, that the management of land is determined by the perceptions, decisions, and actions of millions of users has gradually become the basis for conservation projects and programs.

Incentives and regulations influence land management practices. While regulations involve direct manipulation of quantities, incentives often target relative prices (Chisholm 1987) using voluntary mechanisms to promote the use of land-sustaining practices. Some economists, such as Anders Ekbom in the previous chapter, argue that incentives are likely to be more successful than coercive or legislative measures (Chisholm 1987; Panayotou 2003). A full review of these arguments is beyond the scope of this chapter,[1] but one reason is that information requirements of incentives are likely to be lower. For example, the non-point nature of soil erosion and the high cost of monitoring individual farmers' land management practices make regulatory approaches very challenging to implement—and even more difficult in East African countries, due to weak regulatory capability, high transaction costs, and adverse distributional impacts of regulatory requirements.

Markets may provide some incentives for land conservation. Conservation measures can add to a farmer's bottom line by improving soil quality and productivity, protecting productivity over the long run, conserving water, and so on. Nevertheless, due to the existence of a variety of market failures, additional incentives are used in virtually all conservation programs around the world. For example, when private discount rates are higher than the social discount rate, conservation investments with long time horizons will be depressed. Positive intergenerational externalities from conservation may also not be fully incorporated into private decision making, reducing investments. Similarly, positive spatial externalities from conservation may not be included, also depressing investment below optimal levels.

But underlying factors other than externalities also generate policy imperatives. Imperfections in input and output markets, along with institutional problems such as inappropriate land tenure regimes, poor extension, and low farmer capacity create the need for incentives if land is to be conserved. Hence, in this chapter incentives are defined as inducements for sustainable land management that are designed and implemented by an external body (governments, nongovernmental bodies, or both) to influence the conservation behavior of land users, either individually or collectively, toward a socially optimal conservation goal. They are aimed at making conservation preferable to nonconservation when private and social objectives don't match, resulting in incentives for land degradation.

If conservation practices are privately profitable and land users have incentives to invest optimally, there is little need for the public to intervene. Incentives for sustainable land management are often needed, though, because (a) conservation investment is socially, but not privately profitable; (b) policy and market failures reduce the economic desirability of conservation; and (c) private land users may be unable to invest in profitable conservation practices due to institutional and technical constraints.

In the first case the costs of conservation are not fully recouped, because some of the benefits accrue off-farm due to externalities. In the second, market failures create disincentives for optimal management decisions. A land user may not, for example, be able to get a loan for conservation even though it is a profitable activity. Goods and services markets may also be distorted, leading to underinvestment in land. Finally, land users may be unable to invest due to insufficient information, skills, or capacity.

Types of Incentives

Incentives for sustainable land management can be direct or indirect, based on whether they target land users directly or attempt to change the economic environment within which they operate. Direct incentives are typically designed for specific purposes at the project or program level. They may involve some combination of cash (wages, grants, subsidies) and in-kind payments (food aid, supply of agricultural implements, livestock, trees, seeds, etc.). Direct incentives may also include targeted loans, training, and technical assistance. Indirect incentives are meant to make the economic and institutional environments more conducive for conservation. Indirect incentives involve fiscal measures (e.g., tax relief or input and output price supports), institutional support and services (e.g., extension, technical assistance, training, education, land tenure arrangements, or market development), and assistance to community and farmer organizations (Sanders and Cahill 1999).

Incentives for soil conservation can also be classified as "augmenting" or "enabling." If soil conservation practices are profitable to society but not individuals, a case can be made for society to pay the difference. In this case "augmenting" incentives like subsidies or price supports may be needed. On the other hand, if privately profitable practices are not used because of market imperfections, "enabling" incentives like market improvement, credit supply, technical assistance, extension, and information provision may be required.

Incentives for sustainable land management can therefore be direct and "augmenting" (e.g., grants and subsidies, supply of implements, seeds and seedlings, wages and food aid), direct and "enabling" (e.g., targeted loans or training and technical assistance at the project level), indirect and "augmenting" (e.g., tax relief, price support, land tenure), or indirect and "enabling" (credit services, extension, technical assistance, education and community support). As approaches can vary across countries and regions, a careful assessment of the conditions for intervention is required to determine which combinations of incentives to use.

The mixed success of most incentives for soil conservation in East Africa appears to arise from inappropriate use. For example, in areas where farmers do not have land tenure security it may be futile to expect them to invest in long-term conservation like stone terraces. Further, the use of direct cash or in-kind payments based on the amount of conservation work done may need to be combined with cost-sharing arrangements to avoid the possibility that land users view conservation merely as employment and do not subsequently conserve land (Lundgren 1993). The following four case studies show that, although direct incentives can be quite important in the short- and medium-term, indirect incentives are generally most effective.

Determinants of Improved Land Management Practices and Use of Incentives in East African Countries

After recognizing the severity of soil erosion, colonial powers in Kenya, Uganda, and Tanzania embarked on massive soil conservation schemes. Coercive approaches had farmers construct terraces, sometimes as punishment (Lundgren 1993). Hence, natives associated soil conservation with colonialism: opposition to soil conservation became a focus in the struggle for independence. Leaders such as Jomo Kenyatta of Kenya and Julius Neyerere of Tanzania promised to end soil conservation programs after independence. Indeed, in the 1960s violent opposition to soil conservation programs was common in Kenya, Tanzania, and Uganda (Lundgren 1993).

When colonial-era conservation schemes were abandoned after independence, land degradation worsened as population pressure increased and settlements expanded to marginal areas. Incentives for soil and water conservation and afforestation, mostly with donor support, were therefore widely used in Ethiopia, Kenya, Tanzania, and Uganda (Stahl 1993). For each case I first present a brief summary of the literature on the underlying causes of land degradation and use of incentives. I then analyze whether these incentives have been effective.

Ethiopia

Several studies have explored the relationship between the underlying causes of land degradation and decisions to invest in and use improved land management practices in Ethiopia. Examining a highland setting in central Ethiopia,

Shiferaw and Holden (1998) find that awareness of land degradation is an important factor influencing decisions to invest in conservation technologies, suggesting the importance of education and information programs. The same study strongly posited that conservation that provides short-term benefits to farmers is essential in high population cereal-based systems. Gebremedhin et al. (2003a) investigate the impact of tenure security on investment in the highlands of Tigray in northern Ethiopia and find that land tenure security is an important determinant of investment in stone terraces and tree planting.

In a relatively high potential cereal system in the highlands of central Ethiopia, Shiferaw and Holden (1999) find that poverty increases discount rates and reduces investment. Based on community surveys in the highlands of northern Ethiopia, Pender et al. (2001) find that population growth and frequency of land redistribution (a measure of land tenure insecurity) are associated with reduced use of fallow. The same study finds that credit and technical assistance from microfinance institutions increases use of compost and spurs investment in soil bunds, trees, and live fences. Better market access is also found to have largely positive impacts on land quality. Bekele and Drake (2003), focusing on the eastern Ethiopian highlands, find that financing to cover initial conservation investment costs is positively associated with subsequent household investment in conservation technologies.

Using household, community, and plot-level surveys Gebremedhin and Swinton (2003) find that households in Tigray are more likely to invest in stone terraces and soil bunds on plots they cultivate longer, suggesting the importance of stable tenure (a point reinforced by Kabubo-Mariara and Linderhof in Chapter 4). Households with publicly constructed conservation structures were found to be less likely to invest in either soil bunds or stone terraces, suggesting that public conservation on private land may crowd out private investment. However, the study also finds that nearby conservation on communal lands has positive demonstration effects.

Studies have also looked at the determinants and effectiveness of collective action for community resource management in Ethiopia. Gebremedhin et al. (2004) find that community management of grazing lands contributes to more sustainable use of resources, improves feed availability, and is most effective in areas with intermediate population, high social capital, and low heterogeneity in oxen ownership. In a related study on woodlots, Gebremedhin et al. (2003b) find that external organizations should complement and not replace community resource management efforts.

Use of Incentives for SLM

Ethiopian policymakers largely ignored land degradation until the 1970s. Donor-supported conservation projects were initiated after the disastrous 1973/74 drought, with food provided by the World Food Program (WFP) forming the basis for many conservation programs. The food-for-work (FFW) program started in 1980, and in 1981 a small unit for soil and water conservation in the Ministry of Agriculture was upgraded to a department.

Compensation in the form of FFW and in some cases cash-for-work (CFW) were the main direct incentives for soil conservation in Ethiopia during the 1980s. Under FFW farmers were paid in grain (usually wheat) and edible oil per meter of terrace constructed, numbers of trees planted or seedlings produced, or per day spent on government or NGO-sponsored conservation.[2] The amount of food determined what work was carried out, irrespective of technical suitability. Most was done by farmers who worked to earn food, not to improve the quality of their lands; CFW paid farmers a daily wage rate in cash. In addition to FFW and CFW, tree seedlings were sold at nominal prices for private use and provided free for communities.

Despite a rich indigenous knowledge in Ethiopia, the FFW-based soil conservation programs were aimed at promoting "new" or "improved" soil conservation practices, which were based on little prior research in the country. The programs were fundamentally top-down, with little involvement of local beneficiaries. Moreover, they focused on promoting conservation practices on community lands, with minimal consideration given to individual farms. The fact that food was paid based on the amount of work done made most farmers consider FFW as employment, with little connection to long run soil conservation.

The difficulties encountered by the Ethiopian programs during their initial stage of implementation led to the realization that beneficiaries need to participate in the planning and implementation of conservation programs, including adaptation of conservation technologies to local conditions. Several participatory approaches were used, though the extent of farmer participation and the impact of those approaches on adoption have been limited. With regard to indirect incentives, although extension has included sustainable land management, in practice the focus was on improved crop and livestock production.

The major bottleneck for soil and water conservation in Ethiopia has perhaps been the lack of tenure security. Agricultural land belongs to the state, and farmers have only usufruct rights. Several researchers have documented that insecure land tenure is an important factor inhibiting farmer investments (Gebremedhin and Swinton 2003; Gebremedhin et al. 2003a; Alemu 1998), but until recently no significant efforts were made to increase tenure security. The current government seems to recognize the problem and is conducting land certifications, though the effect of certification on sustainable land management is yet to be seen.

The use of incentives for sustainable land management has changed slightly since 1991. Community mobilization is more prominent than before, especially in the northern highlands of Tigray and Amhara. Although FFW-type programs are continuing, a more bottom-up and participatory approach is being promoted. In addition, community natural resource management, such as area enclosures, community woodlots, and grazing land management has been widely promoted. Although no comprehensive assessment of the effectiveness of community natural resource management efforts has yet been done, case study results indicate positive effects of such indirect incentives (Gebremedhin et al. 2003b; Gebremedhin et al. 2004).

Extension-based education and technical assistance for sustainable land management have expanded significantly since 1991. Every peasant association[3] (PA) is supposed to have one development agent for natural resource management, suggesting that natural resource conservation is becoming an integral component of extension activities. An innovative indirect incentive that has been used since about 1996, especially in the northern Ethiopian highlands, is the distribution of communal degraded lands for private tree plantations. The experience to date on such policies is encouraging, reinforcing the need to improve land tenure security to promote investment in land conservation.

Kenya

Several studies have analyzed farmer conservation decisions in Kenya. Kiome and Stocking (1995), using experimental and survey studies conducted in three field sites in semi-arid Kenya, conclude that farmers' perceptions of land degradation are realistic. Zaal and Oostendorp (2002) conclude that infra-structure and market development, which facilitate the flow of people and information and reduce transport costs, may have encouraged the construc-tion of terraces in Machakos and Kitui districts.

Place et al. (2006a), based on their study in medium- to high-potential Kenya, find that population density and markets promote development of woodlots. Similarly, in their study in central and western Kenya, Place et al. (2006b) find that infrastructure and market development are key factors that differentiate successful intensification of central Kenya from less effective efforts in parts of western Kenya.

Use of Incentives for SLM

After independence in 1963, land degradation was identified by Kenyan authorities as the most severe environmental problem. The National Soil and Water Conservation Program (NSWCP) was started in 1974, supported by funds from the Swedish Embassy (Kimaru and Jama 2006). Gradually it was upgraded into a bilateral conservation program spanning the whole country (Mbegera et al. 1992). Based on the successes of NSWCP, the Regional Soil Conservation Unit (RSCU) of the Swedish International Development Cooperation Agency (Sida) was created in 1982 to promote successful experi-ences throughout East Africa.[4] The core activities of RSCU were training, project initiation and development (Lundgren 1993). Since then Kenya, Uganda, and Tanzania have received support for land management, with RSCU and its successor, the Regional Land Management Unit (RELMA), supporting a variety of projects.

In the mid-1980s the Kenyan government also set up a Soil and Water Conservation Branch in its Ministry of Agriculture, National Environmental Secretariat, and Permanent Presidential Commission on Soil Conservation and Afforestation (Stahl 1993). In 1989 the government established a Ministry for Reclamation and Development of Arid, Semi-arid and Wastelands (Stahl 1993).

The use of incentives to motivate farmers to participate in conservation has changed over the years. At least three developments can be identified (Lundgren 1993). In the beginning subsidies in the form of cash payments were given to people to participate in labor-intensive communal projects like cut-off drains, artificial waterways, and gully control. Employment creation was also an objective of such projects; conservation on private farms was to be done by owners without payment.

In the next phase conservation extension was aimed at individual farmers. Although Kenya initially followed an approach to conservation that emphasized heavy machinery, the country progressively engaged farmers using local resources. The principal incentive was advice and education from extension agents. They then introduced the training and visit (T&V) extension integrated conservation extension system. By 1987, although the focus on individual farmers showed remarkable achievements, it was concluded that the pace was slow and adoption too low in some areas. Hence, the catchment approach was adopted.

Under the catchment approach technicians and soil conservation officers work together with farmers in focal areas to develop land treatment plans acceptable to both farmers and extension agents. The catchment approach focuses on areas of about 50 hectares, which are considered socioeconomically homogeneous. Focal area conservation committees are elected and various farmer groups organized, such as the traditional Mwethya groups of the Ukambani region in Kenya. Catchment conservation committees (CCC) serve as the main means to express farmers' views and put pressure on those farmers who lag behind. Deliberate attention is given to women, with a majority of groups formed and run by women. RELMA activities also must involve women (Kimaru and Jama 2006). By 1993 more than 18,000 agricultural officers were trained in soil and water conservation, and it was reported that more than one million farmers had adopted conservation practices (Stahl 1993). However, about two-thirds of small farms that needed conservation did not have it.

An interesting example of how indirect incentives can be effective can be seen in the districts of Machakos and Kitui. Government support was limited to technical assistance in laying out terraces and occasionally organizing and providing tools to women's self-help (Mwethya) groups involved in conservation (Pagiola 1999). By 1985 about 85% of land in Machakos and Kitui that required conservation had some form on it; on steeper slopes this proportion was as high as 90%.

Pagiola (1999) reports that even with substantial investment requirements and missing credit markets, conservation has been broadly adopted in

Kitui and Machakos without government providing direct incentives. These constraints are, however, alleviated by remittances from family members and significant use of women's labor exchange groups; organizing labor groups has proven to be an effective indirect incentive. Unlike in Ethiopia prior to the early 1990s, Kenya emphasizes indirect incentives like training, technical assistance, and community mobilization. However, it must be noted that despite efforts to improve land management, land degradation is still a critical problem in many parts of Kenya (Kimaru and Jama 2006). The tools used therefore seem inadequate (Bryan and Sutherland 1992) given the infrastructure and market development in much of the country.

Tanzania

Boyd and Turton (2000), based on their study on the Hedaru and Mgwazi villages in the semi-arid plains of the Western Pare lowlands of Tanzania, find that farmers do not adopt soil and water conservation technologies because of both a lack of awareness and a weak extension that rarely includes conservation. They suggest that improved education and information could increase conservation investment. Though most farmers believe they have secure tenure, lack of complementary inputs like tools and seeds limits conservation efforts.

In their study of the North Pare and West Usambara Mountains, Mbaga-Semgalawe and Folmer (2000) find that participation in soil and water conservation programs influences erosion problem perceptions, decisions, and extent of investment. The same study also finds that participation in labor sharing groups (known as *Kiwili* or *Vikwa*) positively influences levels of investment and willingness to use improved soil conservation technologies.

Jones (2002) finds in the Uluguru Mountains of central-eastern Tanzania that knowledge of erosion problems and conservation technologies do not limit adoption. Instead, distance from plots to homesteads and low returns from maize production are the two most important factors. They find that plots have more conservation investment where high-value crops like vegetables are grown.

Use of Incentives for SLM

Several soil conservation technologies were introduced by the British colonial government under the Land Usage Schemes Program started in 1947 (Mbaga-Semgalawe and Folmer 2000). The Schemes were aimed at developing agricultural production systems to rehabilitate eroded areas and generate experience with soil conservation. Local chiefs and agricultural officers were responsible for implementation and enforcing the laws. Protests and riots ensued, and in 1955 the Schemes collapsed.

After independence in 1961, agricultural development and research programs avoided soil conservation. However, partly because areas formerly prohibited from cultivation started to be farmed, by 1970 soil erosion forced Tanzanian authorities to address soil conservation (Misana 1992; Mndeme 1992; Rugumamu 1992). The Hifadhi Ardhi Dodoma (HADO) project in Tanzania was launched in 1973 by the government with funding from Sida (Lundgren 1993). The project focused on rehabilitating badly eroded lands in the semi-arid Dodoma region; as in Kenya, mechanical methods including graders and other machinery were used. Five years in, it was realized that the approach was too expensive and difficult to sustain. The program also poorly incorporated farmers' interests and did not consider socioeconomic realities of the area, ultimately leading to failure.

In 1979–1980, the Tanzanian government—in collaboration with the Regional Integrated Development Program (RIDP) supported by the German aid program GTZ (Gesellschaft für Technische Zusammenarbeit)—initiated the Soil Erosion Control and Agroforestry Program (SECAP) in the West Usambara Mountains. In 1989 the Dutch government initiated an irrigation development program, which included soil and water conservation as a major objective. In 1992 GTZ initiated the Pare Mountain Tanzanian Forest Action Plan (TFAP), with soil conservation as its major component.

The RSCU/RELMA helped to establish the Soil Conservation and Agroforestry Project Arusha (SCAPA) in 1989. SCAPA started as a pilot project, but in 1993 it expanded to cover Arusha and Arumera Districts in northern Tanzania. SCAPA has reportedly adopted participatory approaches to conservation and agro-forestry and provided useful lessons to other projects (Kimaru and Jama 2006). As in Kenya, in a bid to increase farmer participation, groups are formed and farmers trained to plan and construct conservation works using simple equipment. Intensive trainings on nursery management are also given.

Such programs encourage the adoption of soil and water conservation practices in Tanzania by means of various types of incentives. Direct incentives include providing implements and improved seed at subsidized prices. Indirect incentives include technical assistance, revitalization of traditional labor sharing groups to reduce labor shortages, establishment of village-level land use planning committees responsible for conservation, and establishment of tree nurseries. Tanzania is one of the Eastern African countries where indirect incentives are most widely used.

Uganda

Nkonya et al. (2001) examine how soil conservation by-laws are perceived by farmers and how effectively they are enforced. The study finds that, although slash-and-burn restrictions are reasonably enforced, enforcement difficulties mean that rules for cutting trees and burning charcoal are not. Factors positively associated with compliance include land tenure security and the presence of organizations and programs related to resource conservation.

Pender et al. (2006) find that, although households in more densely populated areas and with small farms are more likely to adopt labor-intensive land
management practices, population pressure contributes to soil erosion and
lower crop yields, implying that measures to control population growth may
increase agricultural productivity and reduce land degradation. The same
study argues that land tenure insecurity does not appear to be an issue in
Uganda, since most farmers feel secure. Using a bioeconomic simulation
model, Woelcke et al. (2006) find that indirect incentives like market reforms
and improved credit access contribute to the adoption of conservation practices in Uganda.

Use of Incentives for SLM

Efforts to conserve soil in Uganda started during the colonial period in approximately 1930 (Tukahirwa 1992; Mirro 1999), and chiefs were responsible for
implementing the laws (Zake 1992). Conservation measures included stone
terraces, grass strips, check dams, afforestation and other biological measures.
Between 1940 and 1955 colonial administrators intensified soil conservation using mainly coercion and persuasion, but also incentives like education
and demonstration (Miiro 1999). Soil conservation by-laws were instituted
at the district level in 1956 (Tukahirwa 1992; Zake 1992). Coercive measures
resulted in strong negative attitudes and a sharp decline in conservation after
independence.

Once the severity of the land degradation problem was realized, the Ugandan
authorities took a series of steps to address the problem. In 1986 Uganda established its Ministry of Environmental Protection (MoEP) with a mandate for soil
conservation. In 1990 a National Environment Information Centre (NEIC) was
created, closely followed by the National Environmental Action Plan (NEAP),
the main objective of which was to identify and analyze the major environmental problems and develop a comprehensive national strategy (Sekitoleko
1993). In 1994 the government adopted a National Environmental Policy, one
of the objectives of which was to collect, store, analyze, and disseminate information on the environment. In 1995 the government established the National
Environmental Management Authority (NEMA).

While the establishment of the MoEP provided for unified responsibility
for soil conservation, lack of coordination has been one constraint on effective soil conservation policies (Zake 1992). Other national-level problems
include ineffective extension, lack of appropriate soil conservation technologies, and difficulties in implementing land policies across the diverse tenure
systems (Zake 1992).[5] Although it took some time to recognize the importance of land degradation after independence, a number of soil conservation projects—mostly funded by donors—were implemented. The Uganda
Soil Conservation and Afforestation Pilot Project (USCAPP) was started by
the RSCU and the Ministry of Agriculture, Animal Industry, and Fisheries

(MAAIF) in southwestern Uganda. The project pioneered several useful approaches, including effective farmer participation, community organization, and empowerment and business training (Kimaru and Jama 2006). USCAPP was later scaled up to become the Uganda Land Management Project (ULAMP), covering five districts. In 2001 ULAMP was absorbed by the National Agricultural Advisory Service (NADDS).

Alongside organizational developments, both direct and indirect incentives have been used. Unlike the coercive methods of the colonial era, post-independence approaches rely mostly on extension (Zake 1992). Direct incentives are limited and focus on compensation for labor. Indirect incentives include training, community mobilization, and decentralization of decision-making; these appear to contribute most to land conservation in Uganda. Land tenure insecurity does not seem to be a serious problem, especially in the customary holdings.

Conclusions

A variety of incentives for SLM are used in the East African countries of Ethiopia, Kenya, Tanzania, and Uganda. While Kenya, Tanzania, and Uganda use mainly indirect incentives, direct incentives dominate in Ethiopia. Direct and indirect incentives must be combined judiciously to promote SLM investments and practices. For example, direct incentives could be useful for the introduction and demonstration of conservation technologies, while indirect incentives may encourage adoption and sustainable use.

FFW and CFW programs constructed soil conservation structures and established biological soil conservation in direct attempts to curb soil erosion on private and communal lands. Such approaches failed to recognize the factors underlying land degradation, and they have seen limited effectiveness. On the other hand, indirect incentives generally address the underlying causes of land degradation and are thus more effective in promoting SLM. Since the early 1990s Ethiopia has increased the use of indirect incentives.

The experiences of East Africa point to the need to base SLM interventions on careful analyses of the underlying causes of land degradation. Incentives for soil conservation could be more effective if they were part of larger agricultural and rural development strategies. In areas where land tenure security is an issue, addressing tenure is perhaps the most critical step, because farmers cannot be expected to make long-term conservation investments unless those investments are secure. The low profitability of conservation practices and absence of adequate short-term benefits have also been important barriers to conservation. To encourage conservation at the farm level, incentives like market infrastructure development and price supports that improve the profitability of conservation practices need to be considered. Sustainable management of communal lands also depends on the effectiveness in promoting collective action; indirect incentives such as community mobilization and empowerment are likely to be more effective than direct measures.

Notes

1   For a comprehensive review, see Sterner (2003).
2   About half of all NGOs operating in Ethiopia in the 1980s were involved in soil and water conservation.
3   A peasant association is the lowest administrative unit in the country. It consists of four to five villages.
4   The original geographic mandates of RSCU included Ethiopia, Kenya, and Tanzania, with Zambia included after a few years and Uganda in 1990.
5   For example, customary, freehold, *mailo* (land tenure system where land owners have title to land, but tenants have strong right to land, including eviction protection without compensation), and leasehold systems.

References

Alemu, T. 1998. Farmers' Willingness to Pay for Tenure Security. *Ethiopian Journal of Economics* 7(2): 65–90.

Bekele, W., and L. Drake. 2003. Soil and Water Conservation Decision Behaviour of Subsistence Farmers in the Eastern Highlands of Ethiopia: A Case Study of the Hunde-Lafto Area. *Ecological Economics* 46: 437–51.

Boyd, C., and C. Turton. 2000. The Contribution of Soil and Water Conservation to Sustainable Livelihoods in Semi-arid Areas of sub-Saharan Africa. Oversees Development Institute (ODI) Network Paper No. 102.

Bryan, R., and R. Sutherland. 1992. Accelerated Erosion in a Semi-arid Region: The Baringo District, Kenya. In *Soil Conservation for Survival*, edited by K. Tato and H. Hurni. Marceline, Missouri: Walsworth Publishing Company, Inc.

Chisholm, A. 1987. Abatement of Land Degradation: Regulation versus Economic Incentives. In *Land Degradation: Policies and Problems*, edited by A. Chisholm and R. Dumsday. Cambridge: Cambridge University Press.

Eicher, C. 1994. Zimbabwe's Green Revolution: Preconditions for Replication in Africa. Staff Paper 94–1. East Lansing, MI: Department of Agricultural Economics, Michigan State University.

Gebremedhin, B. 1998. The Economics of Soil and Water Conservation Investment in Tigray, Northern Ethiopia. Unpublished Ph.D. thesis, Department of Agricultural Economics, Michigan State University, East Lansing.

Gebremedhin, B., and S. Swinton. 2003. Investment in Soil Conservation: The Role of Land Tenure Security and Public Programs. *Agricultural Economics* 29: 69–84.

Gebremedhin, B., J. Pender, and S. Ehui. 2003a. Land Tenure and Land Management in the Highlands of Northern Ethiopia. *Ethiopian Journal of Economics* 3(2): 46–3.

Gebremedhin, B., J. Pender, and G. Tesfay. 2003b. Community Natural Resource Management: The Case of Woodlots in Northern Ethiopia. *Environment and Development Economics* 8(1): 129–48.

Gebremedhin, B., J. Pender, and G. Tesfay. 2004. Collective Action for Grazing Land Management in the Highlands of Northern Ethiopia. *Agricultural Systems* 84: 273–90.

Jones, S. 2002. A Framework for Understanding On-farm Environmental Degradation and Constraints to the Adoption of Soil Conservation Measures: Case Studies from Tanzania and Thailand. *World Development* 30(9): 1607–20.

Kimaru, G., and B. Jama. 2006. Improving Land Management in Eastern and Southern Africa: A Review of Practices and Policies. ICRAF Working Paper No. 18. Nairobi, Kenya: World Agroforestry Center.

Kiome, R., and M. Stocking. 1995. Rationality of Farmer Perceptions of Soil Erosion: The Effectiveness of Soil Conservation in Semi-arid Kenya. *Global Environmental Change* 5(4): 281–95.

Lundgren, L. 1993. Twenty Years of Soil Conservation in Eastern Africa. Regional Soil Conservation Unit (RSCU). Swedish International Development Authority (Sida). Report Series 9. Nairobi, Kenya.

Mbaga-Semgalawe, Z., and H. Folmer. 2000. Household Adoption Behaviour of Soil Conservation: The Case of North Pare and West Usambara Mountains of Tanzania. *Land Use Policy* 17: 321–36.

Mbegera, M., A. Erikson, and S. Njoroge. 1992. Soil and Water Conservation Training and Extension: The Kenyan Experience. In *Soil Conservation for Survival*, edited by K. Tato and H. Hurni. Marceline, Missouri: Walsworth Publishing Company, Inc.

Miiro, R. 1999. Factors Enhancing Terrace Use in the Highlands of Kabale District, Uganda. In *Sustaining Global Farm: Selected Papers from the 10th International Soil Conservation Organization (ISCO) meeting held May 24–29 at Purdue University and the USDA-ARS National Soil Erosion Research Laboratory*, edited by D. E. Stott, R. H. Mohtar, and G. C. Steinhardt.

Misana, S. 1992. A Report on Soil Erosion and Conservation in Ismani, Iringa Region, Tanzania. In *Soil Conservation for Survival*, edited by K. Tato and H. Hurni. Marceline, Missouri: Walsworth Publishing Company, Inc.

Mndeme, K. 1992. Combating Soil Erosion in Tanzania: The HADO Experience. In *Soil Conservation for Survival*, edited by K. Tato and H. Hurni. Marceline, Missouri: Walsworth Publishing Company, Inc.

Nkonya, E., R. Babigumira, and R. Walusimbi. 2001. Soil Conservation By-laws: Perceptions and Enforcement among Communities in Uganda. Paper presented at a Workshop on Policies for Improved Land Management in Uganda. June 25–28, 2001, Kampala, Uganda.

Pagiola, S. 1999. Economic Analysis of Incentives for Soil Conservation. In *Incentives in Soil Conservation: From Theory to Practice*, edited by D. Sanders, P. L. Huszar, S. Sombatpanit, and T. Enters. Enfield, New Hampshire: Science Publishers.

Panayotou, T. 2003. Economic Growth and the Environment. Paper presented at the Spring Seminar of the United Nations Economic Commission for Europe. March 3, 2003, Geneva, Switzerland.

Pender, J., B. Gebremedhin, S. Benin, and S. Ehui. 2001. Strategies for Sustainable Agricultural Development in the Ethiopian Highlands. *American Journal of Agricultural Economics* 83(November): 1231–40.

Pender, J., E. Nkonya, P. Jagger, D. Serunkuuma, and H. Sali. 2006. Strategies to Increase Agricultural Productivity and Reduce Land Degradation in Uganda: An Econometric Analysis. In *Strategies for Sustainable Land Management in the East African*

Highlands, edited by J. Pender, F. Place, and S. Ehui. Washington, DC: International Food Policy Research Institute.

Place, F., P. Kristjanson, S. Staal, R. Kruska, T. deWolff, R. Zomer, and E. C. Njuguna. 2006a. Development Pathways in Medium- to High-potential Kenya: A Meso-level Analysis of Agricultural Patterns. In *Strategies for Sustainable Land Management in the East African Highlands*, edited by J. Pender, F. Place, and S. Ehui. Washington, DC: International Food Policy Research Institute.

Place, F., J. Njuki, F. Murithi, and F. Mugo. 2006b. Agricultural Enterprises and Land Management in the Highlands of Kenya. In *Strategies for Sustainable Land Management in the East African Highlands*, edited by J. Pender, F. Place, and S. Ehui. Washington, DC: International Food Policy Research Institute.

Rugumamu, W. 1992. Planning Soil Conservation in Bahi Village, Dodoma District, Tanzania. In *Soil Conservation for Survival*, edited by K. Tato and H. Hurni. Marceline, Missouri: Walsworth Publishing Company, Inc.

Sanders, D., and D. Cahill, 1999. Where Incentives Fit in Soil Conservation Programs. In *Incentives in Soil Conservation: From Theory to Practice*, edited by D. Sanders, P. L. Huszar, S. Sombatpanit, and T. Enters. Enfield, New Hampshire: Science Publishers.

Sekitoleko, V. 1993. Resolution of Conflicts between Agriculture and Environment Protection in Uganda. *Nordic Journal of African Studies* 2(2): 103–08.

Shiferaw, B., and S. Holden. 1998. Resource Degradation and Adoption of Land Conservation Technologies in the Ethiopian Highlands: A Case Study in Andit Tid, North Shewa. *Agricultural Economics* 18(3): 233–48.

———. 1999. Soil Erosion and Smallholders' Conservation Decisions in the Highlands of Ethiopia. *World Development* 27(4): 739–52.

Stahl, M. 1993. Land Degradation in East Africa. *Ambio* 22(8): 505–08.

Sterner, T. 2003. *Policy Instruments for Environmental and Natural Resource Management*. Washington, DC: Resources for the Future.

Tukahirwa, J. 1992. Constraints to Soil Conservation in Uganda. In *Soil Conservation for Survival*, edited by K. Tato and H. Hurni. Marceline, Missouri: Walsworth Publishing Company, Inc.

Woelcke, J., T. Berger, and S. Park. 2006. Sustainable Land Management and Technology Adoption in Eastern Uganda. In *Strategies for Sustainable Land Management in the East African Highlands*, edited by J. Pender, F. Place, and S. Ehui. Washington, DC: International Food Policy Research Institute.

Zaal, F., and R. H. Oostendorp. 2002. Explaining a Miracle: Intensification and the Transition Towards Sustainable Small-scale Agriculture in Dryland Machakos and Kitui Districts, Kenya. *World Development* 30(7): 1271–87.

Zake, J. 1992. Issues Arising from the Decline of Soil Conservation: The Ugandan Example. In *Soil Conservation for Survival*, edited by K. Tato and H. Hurni. Marceline, Missouri: Walsworth Publishing Company, Inc.

CHAPTER 12

Conclusions and Key Lessons

RANDALL A. BLUFFSTONE AND GUNNAR KÖHLIN

I t is time to revisit the huge challenge this book addresses. We and our contributors have focused on the livelihood challenges of over 200 million people
in East Africa, most of whom are members of a group Paul Collier calls the
"Bottom Billion." These are generally people and households who make their
livings through subsistence agriculture in rural areas. Agricultural productivity, growth, and investment in subsistence agriculture are therefore natural
starting points for thinking about what needs to be done to improve the livelihoods of farm households in East Africa.

The scope of this book—agricultural productivity, investment, and sustainability in one of the poorest regions of the world—is wide-ranging but not
comprehensive. While we cover a variety of issues related to sustainable land
management (SLM) and chemical fertilizer applications, the book does not
claim to delve into all that is necessary to demonstrably improve the lives of the
poor. In particular, little or no mention is made of investments in human capital
and common lands, nor do we address the important dimension of public infrastructure such as road, water, sanitation, and energy systems.

Widespread, conservatively defined, well-documented, and astoundingly
stubborn poverty (Deaton 2010) characterize rural areas and summarize the
many micro-challenges facing East Africa. In Ethiopia's Amhara Regional
State, for example, in 2000, 2002, and 2005 over 90% of households had adult
equivalent incomes below $1.00 per day (Bluffstone et al. 2008). Even more
shocking, depending on whether income or consumption is used to measure
poverty, 44% to 64% of households were below the Government of Ethiopia
cutoff for extreme poverty. Such results are confirmed by nationwide studies
(Ministry of Finance 2005), and in 2005 extreme poverty was defined as $0.47
per adult equivalent per day (Bluffstone et al. 2008)!

Improving the performance of agriculture, which as we discussed in
Chapter 1, is the bedrock of livelihoods in rural areas in East Africa and the
entire developing world, is a critical part of reducing the predominance of
poverty and improving livelihoods. Indeed, as pointed out by Morris et al.
(2007), World Bank (2007), de Janvry and Sadoulet (2009), Christiaensen et
al. (2010), and others, a large literature suggests that agricultural growth is

pro-poor. Moving the agricultural sector forward is therefore a critical step toward improving rural household welfare and particularly the livelihoods of the vast majority, who are at the bottom of the world income distribution.

And agricultural output and productivity in sub-Saharan Africa has terribly lagged other regions, with disastrous economic consequences. In many countries agricultural GDP per capita has declined since the 1960s, and in a variety of countries cereal output per person declined (Morris et al. 2007). Cereal yields per hectare have in general increased over time in sub-Saharan Africa, but yields are a fraction of those found in other developing regions, and growth has not been strong enough to keep up with population. This has meant that production has had to increase through expansion of cultivated area, which has particularly impacted forests that provide a variety of use and non-use values in the region. In East Africa, yields in Kenya and Tanzania actually declined by an average of 0.1% per hectare per year (World Bank 2007).

A key reason for this lagging performance has been net loss of soil fertility and particularly loss of soil capital. As discussed in Chapter 1, most of the countries in sub-Saharan Africa have medium to high soil fertility loss of at least 30 kilograms of nutrients lost on average per hectare per year. In three of the four countries considered in this volume, average nutrient loss is over 60 kg/ha/yr, which is much more than fertilizer applications. These data point to the perhaps obvious conclusion that maintaining soil quality and fertility are critical parts of raising agricultural productivity and incomes (Yesuf et al. 2005).

But to preserve and enhance soil fertility and therefore move the agricultural sector forward, investments of a variety of types are needed. This volume examines the necessary investments and discusses factors influencing those decisions. Most authors focus on sustainable land and water management (SLWM), because it is the most important and costly set of investments made by East African farmers. Typically made of earth and/or stone, SLWM structures like earth and stone bunds and water drainage ditches preserve valuable soil fertility, help manage water, and reduce harmful off-site effects.

Such technologies are, however, particularly costly for farmers, a reality that perhaps helps explain their limited use. They are very labor-intensive to construct, with soil bunds requiring 36 to 62 person days per hectare (Ellis-Jones and Tengberg 2000), and terraces even more, requiring 89 to 142 person days per hectare (Mulinge et al. 2010), plus substantial on-going maintenance; moreover, significant agricultural land may be lost to the structures. SLWM investment and maintenance costs are therefore certainly significant from farmers' perspectives. As several chapters have noted, saddling often poor farmers with these costs does not appear to be working. Key contributors and barriers to adoption are therefore evaluated in a number of chapters in this book.

A second critically important technology is fertilizer. As discussed in Chapter 1, chemical fertilizer applications are believed to be a key part of the agricultural output growth observed in developing countries over the last 30 years. Compared with other developing areas, however, chemical fertilizer applications in sub-Saharan Africa and in East Africa in particular are extremely low. While much of Asia averages over 100 kilograms of chemical fertilizer per hectare per year, average applications in Kenya are about 1/3 of that amount and much lower in Ethiopia, Tanzania, and Uganda; understanding why farmers apply so little fertilizer is therefore of substantial importance for increasing productivity (World Bank 2007).

Key Lessons, Findings, and the Huge Challenge

The chapters of this book have examined several themes and offer a number of lessons. Perhaps the most general is that SLWM investment decisions are very complex. For adoption to take place, a series of hurdles need to be overcome. A basic but important complication is that returns to SLWM include a combination of private and public goods. SLWM technologies used in East Africa can reduce erosion and therefore mitigate off-site downstream effects, reducing flooding and improving water quality. The problem, of course, is that off-site benefits are not appropriable by farmers, and we therefore expect underinvestment. Policies to address such underinvestment are the focus of Anders Ekbom in Chapter 10.

Key private SLWM benefits appropriable by farmers may include improved agricultural outputs due to increased soil fertility, soil structure, and water retention. Indeed, this set of private benefits has been the focus of policy outreach activities, because they can be fairly easily cast in terms of agricultural extension. As discussed by Bernanu Gebremedhin in Chapter 11, Kenya and Ethiopia have been reasonably active in incorporating SLWM into extension.

Yet relying on private benefits to spur investments with important off-site effects is certainly not optimal and may not be in any sense sufficient. Recognizing this reality, Kenya has prohibited cultivation on steeply sloped lands, and a number of countries "require" conservation. Given the high rate of soil capital loss, however, it is clear that such rough approaches are insufficient. Understanding how to appropriately spur investments that yield private–public good joint outputs is critical, complicated, and will be discussed below.

A second set of critical takeaway messages from the book is that a variety of factors inform behaviors and affect investments by farmers in East Africa. A promising approach to acquire such insights is productivity analysis, which is essential for understanding the private profitability of investments. Such analysis, done by Menale Kassie in Chapter 8 and Haileselassie Medhin and Gunnar Köhlin in Chapter 9, show that ecological features help determine where SLWM investments are most useful and appropriate. As both chapters discuss with regard to Amhara Regional State in Ethiopia, plots with SLWM structures have lower average yields. Kassie compares these results to the opposite finding for plots in Tigray Regional State and concludes that it is the higher rainfall in Amhara that depresses yields for stone bunds. Indeed, he finds that in high rainfall areas reduced tillage and stone bunds may yield negative private returns, suggesting that environmental factors like rainfall impose boundaries to the appropriateness of SLWM technologies. That reduced tillage and stone bunds are not appropriate under all rainfall scenarios is perhaps not surprising, but that such technologies may be inappropriately promoted is certainly unfortunate. Such findings may help explain low adoption rates of certain SLWM technologies and are important for the design of future interventions.

Medhin and Köhlin use a different technique for analyzing the productivity of SLWM. They find that, while plots with SLWM technologies have systematically lower yields than unconserved ones, treatments tend to be on inherently disadvantaged plots. These plots with steeper slopes and poorer soils are therefore found to produce more than they would have without SLWM; selection of plots by farmers is therefore a critical element. To support extension and interventions for up-scaling, best practices can be identified using the stochastic metafrontier approach to analyze the combination of SLWM technologies, climate, soil, topography, and market characteristics that offers highest levels of efficiency.

But as Shiferaw and other authors discuss in detail, environmental factors are by no means the only ones affecting private returns and adoption. Agricultural policy and institutional factors are also critical and secure land tenure is of special importance; indeed, this finding is perhaps one of the most robust in the literature. A number of authors have found that land tenure is one of the most important determinants of SLWM investments in East Africa (Alemu 1998; Ahuja 1998; Barrett et al. 2002; Gebremedhin and Swinton 2003; Gebremedhin et al. 2003; Shiferaw and Bantilan 2004). The behavior underlying these findings is that farmers who have the assurance that they can capture the returns from costly investments over time are more likely to make such investments. In this volume the importance of tenure is highlighted for Kenya by Wilfred Nyangena in Chapter 3 and Jane Kabubo-Mariara and Vincent Linderhof in Chapter 4, who find that tenure is a key determinant of and positively spurs SLWM investments.

Underlying factors that affect the private profitability of the investments are also important, not least through their impact on farmers' risk exposure. Such factors include poorly functioning product, credit, insurance, and labor markets. As Bekele Shiferaw and Julius Okello point out in Chapter 2, a number of studies find that access to product markets increases agricultural investments and use of inputs (Reardon et al. 1997; Zaal and Oostendorp 2002). That input and output product market access is important is also the main conclusion of the empirical analysis by Fitsum Hagos and Stein Holden in Chapter 6. They find that—adjusting for a variety of factors—accessible product markets are a key determinant of fertilizer use in the northern Ethiopian Tigray Regional State. Nyangena in Chapter 3 finds that accessible markets as an indicator of low transactions costs promotes SLWM investments. Kabubo-Mariara and

Linderhof in Chapter 4 reach similar findings for adoption of terraces; accessible and presumably better functioning product markets therefore appear to increase investments.

As Shiferaw and Okello point out in Chapter 2, evidence regarding the effect of the generally thin and distorted labor markets on agricultural investments is at this point inconclusive. Poorly functioning credit markets, however, appear to depress a variety of agricultural investments. Investments in SLWM and use of agricultural inputs like chemical fertilizers often require credit to finance those purchases due to significant first-cost requirements (Holden et al. 1998; Shiferaw and Holden 2000). But as noted in several chapters, credit markets in rural areas of East Africa are typically very poorly developed and coverage of government-related programs spotty. Households must therefore rely on informal sources of credit (e.g., friends and family), as well as their own resources. Lack of credit is therefore believed to hamper investments (Reardon and Vosti 1995; Holden et al. 1998; Scherr 2000; Swinton and Quiroz 2003).

But channels by which credit market constraints and distortions affect agricultural investments are unlikely to be so unidimensional. Operating in environments of thin labor, credit, and product markets greatly reduces households' degrees of freedom, forcing them to bet more completely on on-farm success. Farmers in East Africa are typically both producers and consumers of their production. When potentially productivity-enhancing but risky expenditures are made, households therefore may be putting their basic livelihoods at risk.

As noted by Mahmud Yesuf and Hailemariam Teklewold in Chapter 5, missing or poorly developed credit markets—and to an even greater extent, missing insurance markets—have important relationships with risk and risk aversion. Though no chapter in the book explicitly examines the relationships between these missing markets and risk, these issues have been extensively examined in the literature (Eswaran and Kotwal 1990; Rosenzweig and Binswanger 1993; Mosley and Verschoor 2005; Dercon and Christiaensen 2007; Yesuf and Bluffstone 2009).

Excessive risk and high levels of risk-averting behavior are treated in this book primarily in terms of responses to risk. As Yesuf and Teklewold discuss, Ethiopian households exhibit some of the highest measured risk aversion in the world, with over half of respondents in parts of Amhara Regional State severely or extremely risk averse. Hagos and Holden find even higher levels of risk aversion in Tigray, with an amazing 89% of respondents severely or extremely risk averse. These figures compare with, for example, a majority of respondents in Zambia exhibiting *intermediate to moderate* risk aversion (Wik and Holden 1998). With such high levels of risk aversion compared with other lower income countries, it is no wonder that underinvestment is such a problem!

Yesuf and Teklewold find that risk aversion—and by extension missing credit, insurance, and other risk-mitigating markets—is negatively related to adoption of chemical fertilizers by farmers. While chemical fertilizers may be mean-increasing, increases in second (variance) and perhaps reductions in the third moments (skewness) of yield distributions may inhibit adoption. These second and third moment effects may be reducing adoption. They also find that household wealth, which often substitutes for credit and insurance markets, is negatively related to respondent risk aversion, but unrelated to fertilizer adoption. They conclude that wealth affects fertilizer use largely through its impacts on risk averting behavior.

This wealth finding is related to those of Nyangena, who finds that social capital likely supports SLWM investments, because of important network effects. Like household wealth, credit, and insurance, social capital is a stock on which households can draw in times of trouble. It therefore also likely substitutes for credit and insurance markets, helps mitigate risk aversion, and, based on the chapter findings, supports investment.

Hagos and Holden also examine the relationship between risk and chemical fertilizer use in northern Ethiopia. They find that risk increases the probability of using fertilizer, but is unrelated to applications. They also explicitly include credit availability through Bureaus of Agriculture microcredit schemes and find households with access to credit are more likely to apply fertilizer and also use more fertilizer than those without credit access.

Salvatore Di Falco and Jean-Paul Chavas take a different tack in Chapter 7 and examine the effects on production risk of seed choice in Tigray. They find that a subtle response to farming in such risky environments is to use many different varieties—and particularly indigenous "landraces"—as hedges against risk. They also highlight some conclusions of the other chapters in Part II by breaking down risk into second and third yield distribution moments. Their research suggests that using more seed varieties actually increases variance in output, but it also increases skewness. Using more varieties is therefore found to reduce the probability of crop failure, which is important enough to counteract variance-increasing effects. These findings are important given the current need to find approaches to reduce the negative implications of climate change on African agriculture. Another conclusion that may be drawn from the chapter is that promotion of fertilizer through "packages" with improved seeds could dramatically increase production risk by reducing crop varieties used.

Key Policy Conclusions

Perhaps hundreds of millions of East African households operate in environments that systematically depress their capabilities and incentives to invest and pull themselves out of poverty. Even without explicitly examining the likely important effects of quasi-public goods provision (e.g., roads, water, energy, etc.) on agricultural investment and productivity, the chapters in this volume identify a series of hurdles and constraints to be overcome in order for investment to take place. We divide these challenges into those related to the profitability of investments and those affecting the public good aspects of

SLWM adoption. The book suggests that governments should work toward alleviating these constraints if agriculture-led growth is to occur.

The most fundamental hurdle to overcome is to make SLWM investments profitable for farmers. For this to happen, technologies need to be well suited to agro-ecological and market conditions, but this obvious fact is unfortunately far from being universally recognized. Factors directly affecting the profitability of investments can be analyzed using standard approaches. These need to be applied to a much greater extent, with more synthesis and analyses, in order to target future interventions and identify technologies for up-scaling.

But looking at factors directly affecting profitability is just skimming the surface. More important for policymakers to understand are the underlying factors that affect investment incentives. Among the most important of these is tenure security. Though Kenya has the most secure tenure in the region, our chapters focusing on Kenya still highlight its importance for SLWM investments. Clearly more needs to be done, because secure tenure is a critical precondition for investment. Recent land certification policy initiatives in Ethiopia are therefore particularly welcome.

Limited options, thin and isolated markets, and risk aversion are highlighted as critical issues in a number of chapters. Any and all efforts to help households hedge and mitigate risk are critical to increasing agricultural investments and output. The literature and results presented in this book are clear that isolated households do not invest, while those with access to product and input markets invest. Reducing isolation through expansion of roads and telecommunication networks is likely to spur investments.[1] The literature on isolation, poverty, and rural economic performance is vast; a full survey is beyond the scope of this chapter. Work in East Africa and elsewhere, however, suggests that access to market towns and roads are very important. Reducing isolation is clearly critical (Dercon and Hoddinott 2005; Dercon et al. 2006).

While provision of insurance in rural areas is perhaps the definitive step toward mitigating risk, insuring substantial portions of East African households currently seems almost a fantasy. Nevertheless, some important pilot insurance schemes have been undertaken—particularly related to health and as part of food safety net programs—and there is perhaps not reason for complete despair (Armendáriz and Morduch 2007, *166–170*; Andersson et al. 2009). Much more important and relevant in the short-run are likely to be increases in credit and banking services that help households save and invest. Such microfinance services have expanded around the world with varying degrees of success. In East Africa experience is still limited in rural areas, but rotating savings and credit association successes in Kenya, Ethiopia, and elsewhere (Nyerere 2004; Armendáriz and Morduch 2007, *57–70*) perhaps support some level of optimism.

But even after mitigating these important direct and underlying agricultural investment disincentives, the critical issue still remains that some investments—particularly related to SLWM—have large positive external effects.

These externalities may be critical to determining whether public intervention is warranted, because the literature is unclear whether even in an environment of complete markets private incentives for investment in SLWM would be sufficient to mitigate the large losses in soil capital observed. Indeed, as discussed in Chapter 1 and elsewhere in this volume, several authors have found that SLWM is often a money-losing proposition for farmers. Although improved market performance and better policies should increase the profitability of SLWM, such changes still might not be enough to actually spur investments. As pointed out by Ekbom, purely private incentives can be expected to lead to suboptimal soil conservation investments.

Instruments for internalizing at least some of the external benefits of SLWM investments are therefore likely to be a critical future policy direction. Providing incentives for economic agents—in East Africa or for that matter anywhere in the world—to invest in public goods is difficult, but internalizing non-point external benefits is, of course, a particular challenge (Segerson, 1988).

Sketching out the instruments needed to reduce soil erosion trods a useful but well-worn path. As of the time of this writing, identifying appropriate classes of instruments is not difficult. But as discussed in Chapters 10 and 11, when we attempt to consider instrument choice and implementation, serious and bedeviling challenges, along with an amazing array of compromises, confront us. As Ekbom notes in his theoretical chapter, fees could internalize off-site externalities, but how to actually implement and collect those fees? In the world of pollution control, when pollution charges are infeasible, polluting products and inputs are often charged (Bluffstone 2003). Such second-best options do not exist when considering the problem of soil erosion in East Africa, because most agricultural production is for subsistence use and therefore nonmarketed. What therefore to tax? Further, given the tens of millions of people who are food insecure in the region it is difficult to envision levying charges on food production, inputs like chemical fertilizers that contribute to downstream effects,[2] or anything else that supports food production.

An emerging literature on instruments for pollution control in developing countries suggests that analysts must not let the best be the enemy of the good, and the good should not be the enemy of the reasonable. As shown for shrimp farming in Thailand (Bluffstone et al. 2006), mobile source air pollution in Ecuador (Jurado and Southgate 1999), and industrial environmental performance in Indonesia (Afsah et al. 2000), creative approaches are required to internalize externalities in developing countries. Such feasible approaches do not obsess about efficiency or even cost-effectiveness, but instead move economic systems toward considering externalities like downstream effects while avoiding huge burdens on public and private sectors (Sterner 2003). Such creativity is clearly required in East African agriculture.

Another way to tackle the problem is through quantity rather than price instruments. Upstream soil conservers certainly perform useful functions for downstream farmers and perhaps others (such as fishermen and other water users) and therefore may deserve payments for the environmental services they provide. In Chapter 10, Ekbom highlights PES, or Payments for Environmental Services, as a particularly interesting compensation policy. He also points out the conditions that make PES appropriate and some critical issues for implementation. This is important, because actually constructing such PES schemes has proven to be surprisingly difficult in East Africa and throughout the world. Indeed, much of the relevant literature is of a conceptual nature (e.g., see Robertson and Wunder 2005). Nevertheless, there is currently an enormous amount of interest in PES schemes[3] and huge possibilities for useful theoretical, empirical, and policy-oriented work that can and should be done.

As a final note, we would like to return to promising recent trends mentioned in Chapter 1. As noted in that chapter, the G8 has committed substantial funds to support sustainable agricultural development (G8 2009) in sub-Saharan Africa, and African governments have agreed to increase public investments in agriculture by a minimum of 10% of national budgets (CAADP 2010). This book suggests that when such programs are implemented they should be done through up-scaling of appropriate technologies identified by scientific analysis. More generally, underlying factors should be addressed and—as Gebremedhin points out—indirect incentives seem to be more effective than direct incentives. We need to continue to sharpen these tools, because the fates of hundreds of millions of poor people are at stake. These individuals in East Africa and other areas of sub-Saharan Africa simply cannot afford to continue the weak agricultural performance observed over the past several decades.

Notes

1   Rural Ethiopia is especially isolated from a telecommunications perspective, with a mobile telephone penetration rate of only 2% compared with over 30% in Kenya and Egypt (Nazret.com 2009).
2   Particularly since the use of chemical fertilizers is likely too little rather than too much.
3   For example, see the Forest Trends and Ecosystem Service Marketplace websites at www.forest-trends.org/ and www.ecosystemmarketplace.com.

References

Afsah, S., A. Blackman, and D. Rutananda. 2000. How do Public Disclosure Pollution Control Programs Work? Evidence from Indonesia. Resources for the Future Discussion Paper 00–44. Washington, DC: Resources for the Future, October.

Ahuja, A. 1998. Land Degradation, Agricultural Productivity, and Common Property: Evidence from Côte d'Ivoire. *Environment and Development* 3: 7–34.

Alemu, T. 1998. Farmers' Willingness to Pay for Tenure Security. *Ethiopian Journal of Economics* 7(2): 65–90.

Andersson, C., A. Mekonnen, and J. Stage. 2009. Impacts of the Productive Safety Net Program in Ethiopia on Livestock and Tree Holdings of Rural Households. EfD Discussion Paper 09–05. Washington DC: Resources for the Future, March.

Armendáriz, B., and J. Morduch. 2007. *The Economics of Microfinance.* Cambridge, MA: MIT Press.

Barrett, C. B., J. Lynam, F. Place, T. Reardon, and A. A. Aboud. 2002. Towards Improved Natural Resource Management in African Agriculture. In *Natural Resource Management in African Agriculture. Understanding and Improving Current Practices*, edited by C. Barrett, F. Place, and A. A. Aboud. Wallingford, UK: CAB International.

Bluffstone, R. 2003. Environmental Taxes in Developing and Transition Economies. *Public Finance and Management* 3(1): 143–175.

Bluffstone, R., D. Anantanasuwong, and I. Ruzicka. 2006. Using Economic Instruments to Address Non-Conventional Pollution Problems in Developing Countries: What to Do about Shrimp Aquaculture in Thailand? *Environment and Development Economics* 11(5): 651–667.

Bluffstone, R., M. Yesuf, B. Bushie, and D. Damite. 2008. Rural Livelihoods, Poverty and Achieving the Millennium Development Goals in Ethiopia. EfD Discussion Paper 08–07. Washington DC: Resources for the Future.

CAADP (Comprehensive Africa Agriculture Development Programme). 2010. www.nepad-caadp.net (accessed January 4, 2010).

Christiaensen, L., L. Demery, and J. Kühl. 2010. The (Evolving) Role of Agriculture in Poverty Reduction. UNU-WIDER Working Paper 2010/36. Helsinki.

Deaton, A. 2010. Instruments, Randomization and Learning about Development. *Journal of Economic Literature* 48(2): 424–455.

de Janvry, A., and E. Sadoulet. 2009. Agricultural Growth and Poverty Reduction: Additional Evidence. The World Bank Research Observer. Advance Access November 9, 2009.

Dercon, S., and L. Christiaensen. 2007. Consumption Risk, Technology Adoption, and Poverty Traps: Evidence from Ethiopia. World Bank Policy Research Working Paper 4257. Washington, DC: World Bank.

Dercon, S., and J. Hoddinott. 2005. Livelihoods, Growth, and Links to Market Towns in 15 Ethiopian Villages. EDRI-ESSP Policy Working Paper No. 3 and IFPRI FCND Discussion Paper #194. Ethiopian Development Research Institute and the International Food Policy Research Institute, October.

Dercon, S., D. Gilligan, J. Hoddinott, and T. Woldehanna. 2006. The Impact of Roads and Agricultural Extension on Crop Income, Consumption, and Poverty in Fifteen Ethiopian Villages. ESSP Conference Paper, International Food Policy Research Institute, June 6–8.

Ellis-Jones, J., and A. Tengberg. 2000. The Impact of Soil and Water Conservation Practices on Soil Productivity: Examples from Kenya, Tanzania, and Uganda. *Land Degradation and Development* 11: 19–36.

Eswaran, M., and A. Kotwal. 1990. Implications of Credit Constraints for Risk Behavior in Less Developed Economies. *Oxford Economic Papers* 42(2): 473–482.

G8 (Group of Eight). 2009. "L'Aquila" Joint Statement on Global Food Security, July 10, 2009. www.g8italia2009.it/static/G8_Allegato/LAquila_Joint_Statement_on_Global_Food_Security%5B1%5D%5D,0.pdf (accessed January 4, 2010).

Gebremedhin, B., and S. Swinton. 2003. Investment in Soil Conservation: The Role of Land Tenure Security and Public Programs. *Agricultural Economics* 29: 69–84.

Gebremedhin, B., J. Pender, and S. Ehui. 2003. Land Tenure and Land Management in the Highlands of Northern Ethiopia. *Ethiopian Journal of Economics* 3(2): 46–63.

Holden, S. T., B. Shiferaw, and M. Wik. 1998. Poverty, Credit Constraints, and Time Preferences: Of Relevance for Environmental Policy? *Environment and Development Economics* 3: 105–30.

Jurado, J., and D. Southgate. 1999. Dealing with Air Pollution in Latin America: The Case of Quito, Ecuador. *Environment and Development Economics* 4(3): 375–388.

Morris, M., V. Kelly, R. Kopicki, and D. Byerlee. 2007. Fertilizer Use in African Agriculture: Lessons Learned and Good Practice Guidelines. Washington, DC: The World Bank.

Mosley, P., and A. Verschoor. 2005. Risk Attitudes and Vicious Circle of Poverty. *The European Journal of Development Research* 17(1): 55–88.

Mulinge, W. M., J. N. Thome, and F. M. Murithi. 2010. Impacts of Long-term Soil and Water Conservation on Agricultural Productivity in Kaiti Catchment, Kenya. In *Agricultural Water Management Interventions Bearing Returns on Investment in Eastern and Southern Africa*, edited by B. M. Mati. IMAWESA Working Paper 17. Nairobi, Kenya: Improved Management of Agricultural Water in Eastern and Southern Africa.

Nazret.com. 2009. "Ethiopia Has Africa's Lowest Mobile Phone Density." http://nazret.com/blog/index.php?title=ethiopia_has_africa_s_lowest_mobile_phon_1&more=1&c=1&tb=1&pb=1. (accessed on December 12, 2010).

Nyrere, J. 2004. The Case of K-Rep: Nairobi, Kenya. In *Scaling up Poverty Reduction: Case Studies in Microfinance*. Proceedings of Conference on Global Learning Proess for Scaling up Poverty Reduction, sponsored by the Consultative Group to Assist the Poor, World Bank, Shanghai, China, May 25–27.

Reardon, T., V. Kelly, E. Crawford, B. Diagana, J. Dione, K. Savadogo, and D. Boughton. 1997. Promoting Sustainable Intensification of and Productivity Growth in Sahel Agriculture after Macroeconomic Policy Reform. *Food Policy* 22: 317–27.

Reardon, T., and S. A. Vosti. 1995. Links between Rural Poverty and Environment in Developing Countries: Asset Categories and Investment Poverty. *World Development* 23: 1495–1503.

Robertson, N., and S. Wunder. 2005. Fresh Tracks in the Forest: Assessing Incipient Payments for Environmental Services Initiatives in Bolivia. www.cifor.cgiar.org/pes/publications/pdf_files/BRobertson0501.pdf (accessed on May 5, 2010).

Rosenzweig, M., and H. Binswanger. 1993. Wealth, Wealth Risk, and the Composition and Profitability of Agricultural Investments. *Economic Journal* 103(416): 56–78.

Segerson, K. 1988. Uncertainty and Incentives for Nonpoint Pollution control. *Journal of Environmental Economics and Management* 15(1): 87–98.

Scherr, S. 2000. A Downward Spiral? Research Evidence on the Relationship between Poverty and Natural Resource Degradation. *Food Policy* 25: 479–98.

Shiferaw, B. and C. Bantilan. 2004. Rural Poverty and Natural Resource Management in Less-Favored areas: Revisiting Challenges and Conceptual Issues. *Journal of Food, Agriculture and Environment* 2(1): 328–339.

Shiferaw, B., C. Bantilan, and S. Wani. 2008. Rethinking Policy and Institutional Imperatives for Integrated Watershed Management: Lessons and Experiences from Semi-Arid India. *Journal of Food, Agriculture and Environment* 6 (2): 370–77.

Shiferaw, B., and S. Holden. 2000. Policy Instruments for Sustainable Land Management: The Case of Highland Smallholders in Ethiopia. *Agricultural Economics* (22): 217–32.

Sterner, T. 2003. *Policy Instruments for Environmental and Natural Resource Management.* Washington, D.C: Resources for the Future Press.

Swinton, S. M., and R. Quiroz. 2003. Is Poverty to Blame for Soil, Pasture, and Forest Degradation in Peru's Altiplano? *World Development* 31(11): 1903–19.

Wik, M., and S. Holden. 1998. Experimental Studies of Peasants' Attitudes Toward Risk in Northern Zambia. *Discussion Paper D–14*, Department of Economics and Social Sciences, Agricultural University of Norway.

World Bank. 2007. *World Development Report 2008: Agriculture for Development.* Washington DC: The World Bank.

Yesuf M., and R. A. Bluffstone. 2009. Poverty, Risk Aversion, and Path Dependence in Low-income Countries: Experimental Evidence from Ethiopia. *American Journal of Agricultural Economics* 91(4): 1022–1037.

Yesuf, M., A. Mekonnen, M. Kassie, and J. Pender. 2005. Cost of Land Degradation in Ethiopia: A Critical Review of Past Studies. Environment for Development Report. www.efdinitiative.org/research/projects/project-repository/economic-sector-work-on-poverty-and-land-degradation-in-ethiopia (accessed on December 12, 2010).

Zaal, F., and R. H. Oostendorp. 2002. Explaining the Miracle: Intensification and the Transition to Towards Sustainable Small-scale Agriculture in Dry Land Machakos and Kitui Districts, Kenya. *World Development* 30: 1271–87.

Index

For Product Safety Concerns and Information please contact our
EU representative GPSR@taylorandfrancis.com Taylor & Francis
Verlag GmbH, Kaufingerstraße 24, 80331 München, Germany